Exploring Perspective Hand Drawing

Fundamentals for Interior Design 2nd Edition

Stephanie M. Sipp

with Cheryl L. Taylor

Publications

SDC Publications
P.O. Box 1334
Mission, KS 66222
913-262-2664
www.SDCpublications.com
Publisher: Stephen Schroff

Examination Copies:
Books received as examination copies are for review purposes only and may not be made available for student use. Resale of examination copies is prohibited.

Electronic Files:
Any electronic files associated with this book are licensed to the original user only. These files may not be transferred to any other party.

Disclaimer:
The author and publisher of this book have used their best efforts in preparing this book. These efforts include the development, research and testing of the material presented. The author and publisher shall not be liable in any event for incidental or consequential damages with, or arising out of, the furnishing, performance, or use of the material.

ISBN-13: 978-1-58503-901-2
ISBN-10: 1-58503-901-2

Printed and bound in the United States of America.

Contents

Acknowledgements

I would like to begin by showing appreciation to an important drawing teacher, Nofa Dixon. She instilled in me confidence to learn to draw. I carry with me the lesson of exploring visual, right brain thinking by filling sketchbook pages with hand drawn images. This provides an opportunity to "noodle around" with the development of ideas for artwork, mind mapping and schematics for projects, classes, and workshops.

I want to recognize my students who continually inspire me to learn and improve my drawing and teaching skills. It is a challenge to create new ways of supporting the learning process by integrating current technology into my methods of teaching. I especially want to remember the women who joined me in creating a drawing community at Martha's Red Door Studio. Our time together provided the spark and encouraged the foundation for this book's focus.

It is a pleasure to acknowledge the participation of many individuals, family members and colleagues who provided assistance, support, and creative input on this project. I am pleased to recognize the unique coaching and support of Jill Johnson, Milanie Hatfield, Ellen McAnany, Patti McClellan, Suzanne Curtis, Helena Helms, Amber Anthony and Karen Bigham.

I am grateful for Lydia Fiser's skilled editing work and layout recommendations. It was a pleasure to work with the team at SDC Publications who assisted in every aspect of this project with encouragement, support and knowledge. A special thank you goes to Cheryl Taylor for her significant contributions including editing, content recommendations and formatting. I appreciate her keeping the bar raised high and her amazing perseverance to the very end.

About The Author

Stephanie Sipp has twenty-five years of learning, creating, and teaching art and design. Originally from New York, she received her Bachelors of Arts and Music from State University of New York in 1977 and her Master's in Interior Design from Florida State University in 1986. She worked in the interior design field for many years before transitioning to the academic arena. She has lived in Jacksonville, Florida since 1992 where she is currently a tenured professor in the Interior Design Technology department at Florida State College of Jacksonville.

Throughout her career, she has also been exploring art and has created a collection of original artwork using line drawing, watercolor, acrylic, and collage mediums. Working within a lifetime interest in the outdoors, healing gardens and bird watching, her art expression celebrates the diversity and power of nature.

Stephanie's dedication to promoting and teaching the visual arts is evident in her artwork and her facilitating. She brings her interest in the process of creativity to her teaching, encouraging personal expression through the arts. Outside the college classroom, she has developed and facilitated numerous seminars, design principles workshops, beginning drawing classes and continuing education units for professional designers, focusing on art, creativity, self-expression, diversity, and team building. Within the Interior Design department, her expertise includes hand sketching, perspective drawing and color rendering with markers and pencil. She encourages all of her students to meet the challenge of learning to draw and understands that drawing is a learned skill that develops with proper guidance and practice.

Preface

I was not one of the rare individuals born with the innate ability to draw. It has evolved with many years of learning, studying, and practice; and, I am still learning and developing my skills. As an interior design professor, I found that many of my students came with little knowledge of how to draw and with a great deal of self-doubt. I also found that many of the available drawing books were either too advanced or too broad and with little instruction for drawing interior elements.

This book introduces basic line drawing with attention on interior elements. It focuses on simple methods for learning beginning drawing skills and provides encouragement for overcoming typical obstacles. I start from the most basic concepts and then provide progressive systematic instruction by taking students through a series of activities in the supplemental *Activities Sketch Book*. Each activity builds upon the previous with the goal of producing complete one-point and two-point perspective rooms.

I believe the skill of hand drawing, whether with pencil and paper or on an iPad, is an essential tool in the design process. With advancements in computer technology designers can now quickly create complex and precise drawings. However, many will advocate that hand drawing enhances the understanding of fundamental spatial concepts necessary to help the designer and artist translate a visual idea. For me, formulating ideas for new projects starts with hand sketches, symbols, words and graphic notes. The hand drawing time is a slower process providing a longer time for contemplation without the technical distractions of computer software.

All of the drawings in this book are hand-drawn and reflect an element of imperfection and personal style. This approach is intentional so that students do not fall into a trap of expecting their images to look as precise as computer generated images.

It is my hope that this book will open a door of new possibilities on your journey into drawing and exploring your creativity potential.

Part I
Fundamentals of
Drawing

The role of drawing in design is to record what exists and to work out ideas, speculating and planning. Drawings help guide the design development process from idea concept to proposal to construction. This book combines systematic instructions, hand drawn illustrations and video demonstrations, to provide every student the opportunity to learn hand line drawing.

Part I focuses on techniques for learning fundamental hand drawing skills. It covers basic tools and techniques, elements and principles of design, and the creation of basic forms by adding value, texture and pattern, shade, and shadow.

Chapter 1

Orientation

As with any new project, it is important to have the right tools and supplies available before starting the work.

In this first chapter, we will begin by putting together a set of essential supplies and then review a few basic concepts. We will then start putting pencil to paper, which can sometimes be very intimidating if you have convinced yourself that you cannot draw.

You too can draw! So, let's get started.

GETTING STARTED

Learning to draw is no different from learning to play an instrument, learning a new sport, or learning how to cook. The critical components include instruction and practice. When learning to play a musical instrument most people start by taking lessons, practicing scales and playing simple compositions before moving on to more complicated pieces of music. Think about learning to play a new sport. Again, taking lessons, learning to use the equipment and practicing key moves will bring you better success when you play the game. Finally, each individual learns in their own time frame and accomplishes different levels of skill.

The companion *Activities Sketch Book* provides systematic activities for you to complete as suggestions for practicing this skill. I encourage you to repeat any activity that you want to improve upon. In addition, I encourage you to adjust or change the activity to meet your own drawing practice needs.

While some activities have examples of objects that I have drawn, I recommend you find your own object to draw. Copying my image will not provide you with the same practice as drawing from an object in front of you.

Leonardo Da Vinci was a genius who I really appreciate and admire. He completely covered his sketchbook with drawings as he worked out ideas and practiced his drawings. To create a "Leonardo Da Vinci page", try practicing an activity several times. Use different leads or add your drawing marker. The goal is to improve your skills. Be willing to do what it takes to get there.

A page from my sketchbook

DRAWING TOOLS

When you are beginning, it is critical to become familiar and comfortable with your supplies, so keep them simple. Having a favorite pencil, drawing marker, eraser and ruler are more important than having many different choices. You can become distracted easily by wanting to use different supplies and failing to know them all very well.

Having fewer supplies also makes it easier to contain them in a small portable pouch. This gives you an opportunity to carry them with you during the day. The supplies are easily accessible to you while you are waiting in the car, at the doctor's office, or anywhere that you can draw. There are opportunities to pick up 15, 20 or 30 minutes of time to practice. It is great to have your supplies at hand when you have these little snippets of time.

BASIC DRAWING SUPPLIES

Drawing Pencils	HB, B & 2B wood drawing pencils one each
Eraser	White block eraser
Sharpener	Hand held, small pencil sharpener
Drawing Markers	Thin, medium, wide drawing markers in black or sepia
Eraser Shield	Small metal tool with openings used with erasing
"T" Square	12" small, plastic "T" square
Triangle	90 degree small drafting triangle

BASIC DRAFTING TOOLS

In Chapters 8 and 9, we will be creating large one-point and two-point perspective grids and rooms. In addition to your basic drawing supplies, you will need your basic drafting tools to complete the projects. These additional supplies include for following:

Roll of Tracing Paper	Least expensive, white or yellow 24" wide tracing paper
Drafting Tape	Low adhesive drafting tape or dots
Architectural Scale	Triangular shaped or flat architectural scale
Triangle	A 8" or larger architectural triangle
T-square 36 inch	Metal or wood 36 inch T-square
Drafting Surface	Drafting surface large enough for a 24" X 36" paper
Velum Paper	Plain 24" X 30" velum paper

PENCIL LEAD

As noted in the drawing supply list, I recommended you start with an HB, B & 2B pencil lead. Drawing lead comes in a variety of qualities ranging from soft (B) to hard (H). This provides you with options depending on the goal for your drawing. Also, everyone applies a different strength as they push their pencil across the page. You will need to experiment to find which lead works best for you.

Lead comes in 13 different forms ranging from 6B to 6H, with the middle point being called HB. The scale below shows the various lead options.

6B, 5B, 4B, 3B, 2B, B, HB, H, 2H, 3H, 4H, 5H, 6H

Soft...............................Mid............................. Hard

When drawing freehand, having a soft lead will assist with the fluidity of your drawing. It is easier to draw a round circle with a 2B lead since the softer lead will move more easily around the paper. As you move into 3B lead or higher the graphite is so soft it will leave too much smudging on your paper. When hand drafting, the goal is to have a consistent line diameter; using a mechanical pencil and lead will accomplish this objective.

DRAWING MARKERS

As you get more comfortable drawing, using a black drawing marker will be important. Drawing markers are made by several different manufactures. In addition, they come in a range of very thin, thin, medium and brush tips widths. It will be important for you to explore these options and find the drawing markers you prefer. You will want to use black instead of colors.

TRACING PAPER AS A TOOL

One of the best tools you can use is tracing paper. Here are two ways that I find it useful:

❶ **Using tracing paper to learn to draw an object.**

Part of learning a new skill is teaching your body what to do. As one of my students said: "I know what I want to draw, the problem is the distance from my mind to my arm is too long".

The more you draw, the closer your drawing will be to what you want. You can start by using tracing paper to learn to draw an object. Simply lay tracing paper over an image and then copy the drawing. This will help give you the practice so eventually your hand will draw what your mind wants it to draw. Throughout the book, there are basic drawings that you can lay tracing paper over to practice drawing in this manner.

❷ **Using tracing paper to refine a drawing.**

You can also use tracing paper to refine a drawing. One way to do this is to draw the box, then lay tracing paper over it and redraw the object using only the lines you need. For example, you can draw the back of the box to be sure the shape is correct, but when you refine the object, do not redraw the back. You can use as many tracing paper overlays as you need while you are refining a drawing.

TRIANGLE AS A TOOL

Using a triangle as a tool assists with getting the perspective lines matched with vanishing points. You can put your pencil point on the vanishing point, lean the triangle on the pencil, and align the edge of your triangle with the corner of your object to help draw straight and properly angled perspective lines.

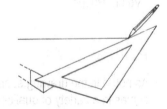

12" PLACTIC T-SQUARE

The small plastic T-square can be used as a straight edge and can help create 90-degree angles. By leaning the "T" portion of this tool against the sketchbook edge, you can draw a straight line across the page, either vertically or horizontally. Use this technique for basic guidelines and beginning pencil drawings. You will need this tool less frequently as you become more confident with drawing straight lines.

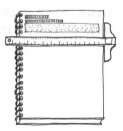

KEY CONCEPTS

Here are some key drawing technique concepts used throughout the book. With each chapter, we will add new concepts.

Line. A basic design element and used throughout the book.

Consistency with line. The quality of keeping your drawn line the same thickness.

Looking ahead. As you draw your line, you will have your eye look ahead to the next dot or line.

Guide points. Light pencil dots made on your page to provide a starting and stopping place.

Guidelines. Preliminary light pencil lines that can be adjusted easily before adding a darker pencil line or drawing marker.

Line to line. The technique of touching each line to another line.

Line weight. The thickness of a line. Usually the outside lines of the object are thicker; while the inside detail lines that create texture and pattern are thinner.

Tracing paper overlay. Using transparent paper as an overly on a drawing to retrace and refine the drawing.

Practice. A key aspect of learning a new skill by repeating activities repeatedly until you are comfortable with the lesson.

THE LINE

The Line is the basic component in this book. The goal is to use a consistent line to draw the outline of an object and to add other elements such as texture, pattern, shape, shade and shadow.

To begin drawing a consistent line, one that has the same thickness, start your line on one point then move your hand across the page to another point in one stroke. You should also practice looking ahead by moving your eye to look beyond the position of your pencil. Looking ahead towards or at your target point will help you create a more consistent, straighter freehand line.

This process starts with **guide points** drawn on the page. Start with a point on the left and a second point on the right. Put your pencil on the left point, look ahead to the second point, and move your

Avoid this start and stop type line.

pencil across, keeping your pencil on the paper the whole time. Avoid the stopping and starting motion. Practice until you can make a confident, smooth single line across the page.

Another technique of supporting your drawings is to use **guidelines**, or light pencil lines used in the early drawing stage. You can use these preliminary lines to check your shape and proportion. You can easily adjust these lines before adding your darker pencil or drawing marker lines. In later chapters, we will start using line to depict texture, pattern, shade, shadow and shape.

ELEMENTS & PRINCIPLES OF DESIGN

Combining drawing techniques with the elements and principles of design will provide a starting point for practicing your drawing skills.

Here are some key elements and principle design concepts introduced in this chapter.

Elements. The parts, or components, of a design.

Emphasis. The main element or focal point; what the viewer's eye should see first.

Geometric shape. usually shapes that are precise and exact. Examples are triangles, squares and circles.

Negative space. A background area without content that supports overall composition.

Pattern. A motif that creates an orderly whole.

Positive shape. A shape or line placed in a negative or empty space.

Principles. Ways the parts or elements are used, arranged, or manipulated to create the composition of the design; how to use the parts.

Proportion. The relative measurements of dimensions of parts or a portion of the whole.

Shape. An image in space.

Space. An empty, negative area where our design will fit.

Texture. The surface quality of an object.

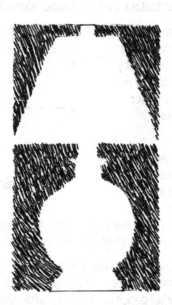

SPACE & CONTRAST

An important quality of your drawings is visual communication of your object as a figure. Objects become more apparent when viewed against a contrasting background. In this example, contrast of figure shape to background creates the lamp image.

NEGATIVE & POSITIVE

The background is considered the negative shape, while the objects, such as the lamps, are considered positive shapes.

LINE & SHAPE

Line is a basic graphic element of drawing. It is a linear mark that shows and describes the contour of an object. These contour lines provide the shape and general nature of the object.

Shape is an image in space. Simple geometric shapes, like a square, can be used as a tool for drawing objects in proportion. In this activity, the square shape provides a method for finding the center of the shape and as a starting shape for drawing circles.

Many of the objects that you will be drawing start with a simple shape. Here is a series of drawings that start with a basic shape to create furniture shapes in a plan view. The plan view represents a projection of shapes when looking down on the floor plan.

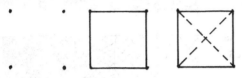

Row 1 – Creating a box starts with 4 guide points that are connected by joining lines. A dashed X from corner to corner marks the center and acts as guide lines for adding additional detail.

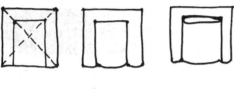

Row 2 –The dashed lines are used as guides to draw the inside lines forming the arms of a chair. The last drawing has an added pillow shape.

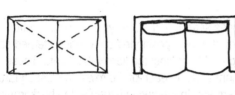

Row 3 – Expand the box to a rectangular shape to form a sofa shape. The dashed X again marks the center of the sofa.

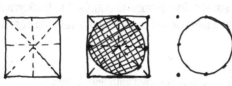

Row 4 – A square shape starts the round table shape. Dashed guideline "X" placed from corner to corner and a cross through the center. The addition of guide dots to the cross aid in drawing the circle.

Drawing of the furniture floor plan used the steps above. The shapes were refined to represent each furniture piece. Suggestion of a lamp was added to the round table and suggestion of a plant added to the rectangular coffee table. For the final image, drawing marker was added and the pencil lines were erased.

VALUE & PATTERN

The design elements of *value* and *pattern* are important components when drawing objects and communicating your designs ideas for interior design. These two elements are used together to provide visual variety and contrast.

Value refers to the range of possible lightness or darkness achieved with a marker or pencil. Gradations moving in degrees from light to dark express value, as seen in this value study.

Pattern is the arrangement of forms or shapes that create an orderly whole. There is usually a variety of pattern in an interior space. A variety of geometric patterns is used in this gradated value study by using crossing lines, diagonal lines, circles and zig-zag patterns.

VALUE CONTRAST

Contrasting value is a method for visually distinguishing shapes seen together. A typical example when making interior design selections would be using dark patterned pillows positioned on a light patterned sofa. The darker valued pattern pillows become more easily recognizable against the light valued sofa. These illustrations below show high, medium and very high value contrasts. In each of the rows, the first square has high contrast with a dark background and a light center square. In the second square, the background and center square are similar in value. In the last square, the lighter background and the very dark center creates very high value contrast.

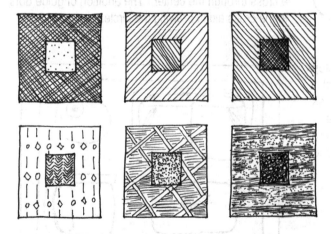

Value contrast with line & dots.

Value contrast with pattern.

LINE WEIGHT

Line weight provides width or thickness of line. Design drawing uses a typical formula for using line weight as a method for showing aspects of an object. The chair image below demonstrates how the four different line types are used and how each line type communicates an aspect of the chair design.

A – Object Edge Line

The heaviest solid line that defines the shape and profile of the object. These contour lines are important because they are the edge of the solid mater separating the object from others.

B – Inner Object Line

The medium solid line that shows the interior contours that depict the three dimensional volume.

C – Surface Line

The lightest solid line that indicates a change in color, value or texture on the surface of the object.

D – Hidden Line

The lines that show edges not usually seen and assist in communicating structure.

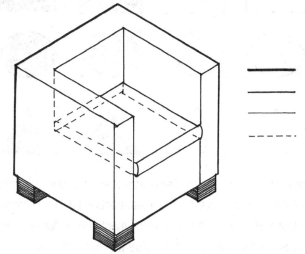

Helpful Hint

When drawing lines that need to be different widths, it is best to have drawing markers that are in width sizes. These marker widths can range from very small (0.1, to very thick (0.8). They come with different qualities such as a metal or plastic tip, pigmented ink, archival and acid-free. These sizes can vary depending on the manufacturer. It is best to purchase several sizes from several different manufacturers to find the pen type and nib size that you like to work with best. The drawing illustration below was drawn with Staedtler, pigment liner pens.

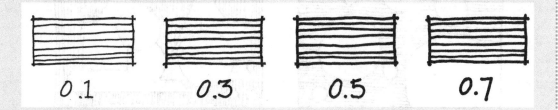

PUTTING IT TOGETHER

Shapes are the basic components of the objects that you will be drawing. Here is a series of drawings that all started with a square. Guide points and lines were used to assist with this process.

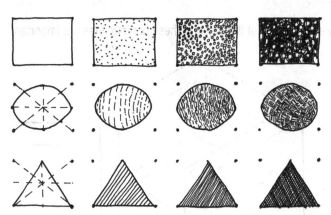

A simple way to start drawing is to use guide points and square shapes. This method was applied to create each of these shapes.

Pattern was added with gradated value.
- Row 1 – pattern of circles
- Row 2 – pattern of broken lines
- Row 3 – pattern of lines

The four rectangles below demonstrate the elements and principles of design:

A - Circle shapes are the positive element and they are drawn in a variety of sizes.

B - Circle shapes are the positive element, the circles overlap, drawn with a variety of size and patterns.

C - Negative space has a dark value with circle shapes.

D - Variety of shapes, circle, square, triangle and rectangle are used with a contrast of dark negative space.

Creativity Strategy
FINDING THE TIME TO DRAW

With all the demands in life, carving out quality time to work on your drawing skills can be a challenge. However, I have found a few strategies that help me find the time to practice. Here are the steps I take that may help you.

1 Find your creative energy zone. Think about the time of day where you can bring your best creative energy and are not tired. Review your week and note what you have planned during that optimal creative-energy time. Maybe you can alter your schedule to allow for drawing during your most creative part of the day and schedule other activities, like shopping and chores, for times when you are tired.

2 Make an appointment with yourself. Once you have identified your creative-energy time, write in appointments on your calendar that say "Working on Drawing." If you block out specific times in your week to draw, you are more likely to actually practice.

3 Redefine your drawing time. Another helpful tip is to tell people you are "working". Unfortunately, our culture respects *working* more than *creating*. Therefore, if you present it to others as work they are less likely to try to convince you to do something else. This is especially true with homework.

4 Keep it manageable. I have found that scheduling each drawing appointment for a half hour to an hour is easier to fit into my schedule. If you schedule three one-hour drawing appointments a week and stick to this for a year, you will have spent 156 hours on drawing.

5 Fill voids with drawing practice time. If you bring your sketchbook and drawing tools with you, you can find spare moments to practice drawing -in the doctor's office, riding in the car, sitting in that boring meeting.

I did the drawing on the right while I was attending a three-hour meeting. This was the furniture piece at the front of the room. Although it is not an outstanding piece of furniture, it provided me with a subject to practice my skills. Remember to look carefully at what you're drawing to be successful.

Chapter 2

The Box

The box will be our starting point for learning to draw any object and will provide a method for adding proportion and understanding perspective.

Creating proper proportion and perspective are two of the most challenging aspects of developing successful drawings. The box method helps us achieve both of these goals.

Throughout this book, we will work with the box. We will move the box around, create objects from the box, divide the box, and embellish the box.

Ultimately, we will transform our boxes into entire perspective drawings of interior rooms, complete with furniture and accessories.

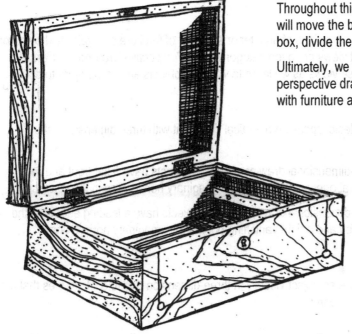

GETTING STARTED

The following are concepts that will help you throughout Chapter 2:

Proportion. The correct relationship of size between two or more parts of an object.

Perspective. The appearance of a distant object in relation to the observer's distance from it.

Perspective lines. Lines that project from the front of the object towards a single vanishing point

Horizon line. Imaginary line that is at eye level on a perspective drawing.

Horizontal line. Lines of the object that are parallel to the horizon.

Vanishing point. The point where parallel lines converge in a perspective drawing.

Vertical lines. Lines of the object that are perpendicular to the horizontal lines.

Leading edge line. The front vertical line of the box when the corner of the box is facing you

Parallel lines. Lines that remain the same distant apart and never meet.

Perpendicular lines. Lines that meet to make a right angle (90-degree).

Scale. A ratio representing the size on a drawing. Typically, architectural scales include 1/4 inch and 1/2 inch equals one foot.

VIEWPOINTS

To communicate a design concept fully, drawings of an interior space or of furniture are typically drawn from different points of view. These different views are often categorized as *Perspective* Drawings *or Paraline* Drawings. We will use a subset of these views as we learn to visualize objects and as we learn to draw.

PERSPECTIVE DRAWINGS

One and two-point perspective views depict space on a vertical plane but with three dimensions, thus creating a more natural view.

One-point perspective views are three-dimensional drawings in which objects have a flat front and the parallel lines depicting depth converge at one single point on an imaginary horizon line in the distance.

Two-point perspective views are three-dimensional drawings where objects have a leading edge and the horizontal parallel lines converge at a left and a right vanishing point on an imaginary horizon line.

PARALINE DRAWINGS

Paraline views provide a means describe an object visually in three-dimension *and* to scale. Lines that are parallel in reality remain parallel in the drawing.

The following drawings of the box shown in the photograph represent the different views and aspects of the box design.

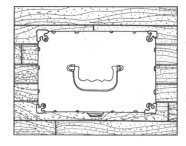

This is the *top view* of the box. You are looking directly down on the box and there is no depth shown in the handle. You will see the details of the box top and the wood flooring. This drawing was done using a scale.

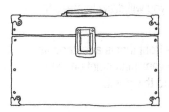

This front view shows detail of the object in an *elevation view*. Elevation drawings do not show depth. This drawing was done using a scale.

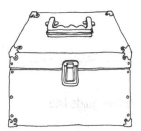

A *one-point perspective view* provides a three-dimensional depiction of the box. Notice the flat front of the box and the parallel lines forming the sides of the top recede toward a single point. The proportion was drawn without a scale.

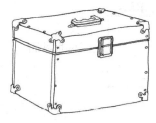

A *two-point perspective view* also depicts a three-dimensional aspect of the box. In this view, there is a leading edge and each side appears to get smaller as they move away from the leading edge. The proportion was drawn without a scale.

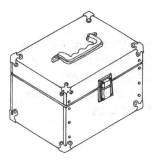

This *paraline* drawing shows the three dimensional qualities of the box. However, it is distinctly different from a perspective drawing because parallel lines do not converge and a scale was used to complete the drawing.

ONE-POINT PERSPECTIVE

FLAT FRONT BOXES

Drawing an object in perspective provides a realistic view and is therefore an important type of drawing for the interior designer. To view a box in a *one-point perspective*, hold it with a flat front facing you and so the two sides appear to move toward a single point in the distance.

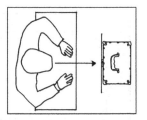

The surface area of the picture, or the *picture plane*, is often thought of as a window through which you are seeing the three dimensional object. In the one-point perspective, the picture plan is perpendicular to your view of the object and the object has a flat front.

With a one–point perspective box, or a flat-front box, the three types of lines you will draw to make the box include horizontal, vertical, and perspective lines.

Notice the boxes drawn below and the use of these lines. They each have a front that is a rectangular shape and the lines defining the side edges of the box move toward a single vanishing point on the horizon line. The vanishing point is stationary and the perspective lines project toward this point.

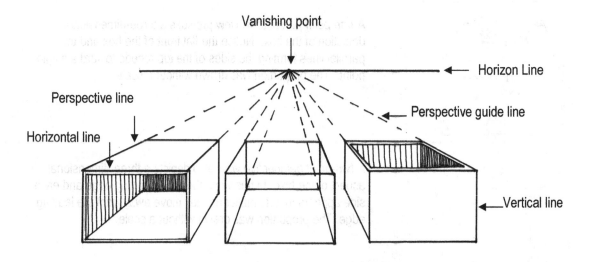

Vanishing point

Horizon Line

Perspective line

Perspective guide line

Horizontal line

Vertical line

Helpful Hint:

Perspective drawings are different from scale drawings because the scale drawing uses measurements to note size. A perspective drawing represents how our eyes see the world naturally, which is not in a measured scale. Objects are scaled relative to the viewer. Additionally, an object is often scaled unevenly: a circle often appears as an ellipse, a square can appear as a trapezoid, and objects appear to get smaller as their distance from the viewer increases. This distortion is referred to as foreshortening and is a technique that helps create the illusion of depth.

ONE-POINT PERSPECTIVE BOX STEP-BY-STEP

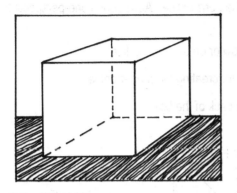

Throughout this book, the starting point for drawings objects and furniture pieces is the box shape. The transparent box provides a way to see the volume of the shape. In this section, we will detail a series of steps to create the one-point perspective box. In this example, we will use a box that is below the eye level or horizon line. The box is drawn with three types of line - a parallel line, a perpendicular line, and angled lines from the vanishing point. The dashed lines in these illustrations are the perspective guidelines.

Drawing Visible Edges:

① The horizon line (HL) is the solid line on top with the single vanishing point (VP) on the far right side.

② Draw the flat front square box below the horizontal (HL) making the top lines parallel to the horizon line and the vertical lines perpendicular.

③ Starting at the outside top corner, draw dashed lines to the vanishing point with a straight edge tool. Repeat this for the other two corners.

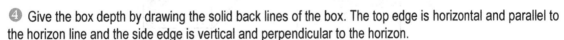

④ Give the box depth by drawing the solid back lines of the box. The top edge is horizontal and parallel to the horizon line and the side edge is vertical and perpendicular to the horizon.

Drawing Hidden Edges:

① Starting at the bottom left front corner, draw a dashed line to the vanishing point.

② Draw a solid line from the back right corner, parallel to the horizon line until it meets the dashed perspective guideline.

③ To finish the box shape, where the back box line meets the dashed perspective guideline, draw the vertical line, perpendicular to the horizontal line (HL)

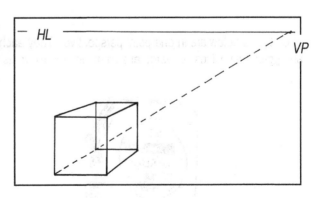

CREATING ONE-POINT PERSPECTIVE OBJECTS

In the drawing below, objects are at varying points relative to the horizon line. Again, each one-perspective box was created as follows:

❶ Start by drawing a single flat rectangular shape above, below or on the horizon line.

❷ Add perspective lines using the vanishing point as a guide for creating the proper angle.

❸ Add additional horizontal and vertical lines to complete the back of the box.

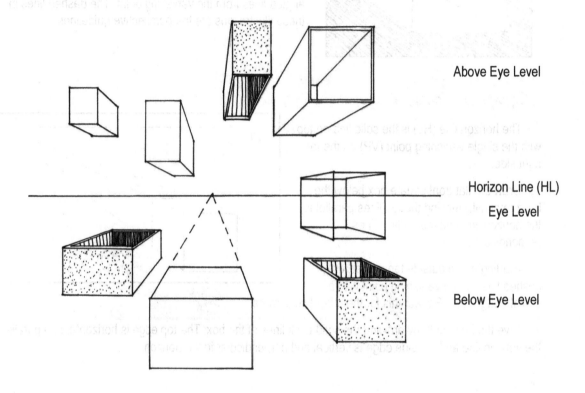

Above Eye Level

Horizon Line (HL)

Eye Level

Below Eye Level

The objects below are in one-point perspective. They each have a flat front and perspective lines converge at a single imaginary vanishing point.

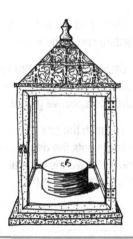

TWO-POINT PERSPECTIVE

LEADING EDGE BOXES

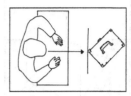

Another type of design perspective is the *two-point perspective*. In this view, we will turn the box so the corner is facing you, which creates a *leading edge*. *Perspective lines* project and converge at two different points on the *horizon line*. In a two-point perspective drawing, the object is placed on a 45-degree angle and the leading edge is on the picture plane. Also, only vertical lines and perspective lines are used to draw the box.

Notice how *perspective lines,* those that make up the top and sides, project towards two separate vanishing points on the horizon line.

The remaining *vertical lines* complete the shape of the object and are parallel to the leading edge.

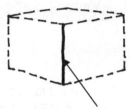

Leading Edge

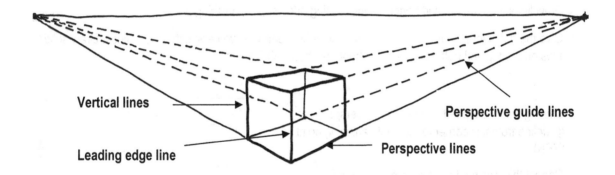

Vertical lines

Leading edge line

Perspective guide lines

Perspective lines

Helpful Hint:

One of the challenges to drawing boxes or objects in two-point perspective in your activity book is that the vanishing point locations on your paper are very close together causing the drawing of a box shape to look unrealistic and distorted. This is because, for the purpose of learning the two-point perspective concept, we are using two vanishing points on a piece of paper that are actually too close together. In reality, the correct vanishing points are about 6 feet apart. Imaginary vanishing points and guidelines are used as you become more familiar with the drawing steps.

TWO-POINT PERSPECTIVE STEP-BY-STEP

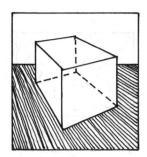

As we did with a one- point perspective box, we will detail the steps for drawing the transparent two-point perspective box. Again, we will start with an objects positioned below the horizon line. The dashed lines represent perspective guidelines.

Drawing the Front Edges

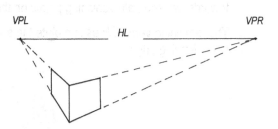

❶ Draw the horizon line (HL) with two vanishing points - one on the left (VPL) and the other on the right (VPR).

❷ Draw the leading edge - a line that is perpendicular to the horizon line. The length of this line will determine the size of the box.

❸ Draw lines from the top and bottom of the leading edge to the vanishing points.

❹ Give the box depth by drawing solid vertical lines between the two perspective guidelines. Draw solid lines on the top and bottom of the box shape as shown above.

Drawing the Top of the Box:

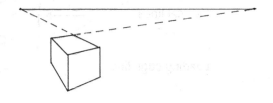

❶ Starting with the top left corner, draw a perspective guideline from this corner to the right vanishing point (VPR).

❷ Repeat this with the top right corner to the left vanishing point (VPL)

❸ Where the perspective guidelines intersect will be the top of the box. Draw solid lines representing the box top shape as shown.

Drawing the Back of the Box:

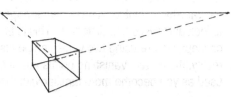

❶ Starting with the lower left corner, draw a perspective guideline from this corner to the right vanishing point (VPR).

❷ Repeat this with the bottom right corner.

❸ Where the perspective guidelines intersect will be the bottom of the box. Draw solid lines representing the bottom of the box shape as shown.

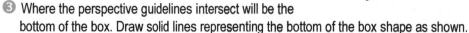

CREATING TWO-POINT PERSPECTIVE OBJECTS

In the drawing below, objects are at varying points relative to the horizon line. The boxes were created using the following steps:

① Mark a left and a right point on the horizon line to denote the vanishing points used to establish perspective.

② Draw a vertical line anywhere above, below, or over the horizon line. This will define the leading edge of your box.

③ Add perspective lines starting from the top and bottom of the leading edge line. These lines are angled so that they are directed towards either the left or the right vanishing point.

④ Add additional vertical lines to complete the back edges of the box.

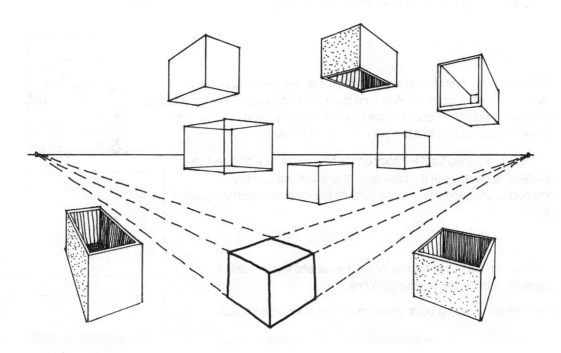

The objects below are in two-point perspective. They each have a leading edge with perspective lines that converge at imaginary vanishing points.

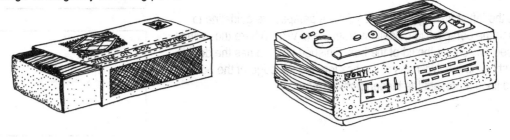

OTHER GEOMETRIC SHAPES

The transparent box shape can be a starting point for drawing other geometric shapes.

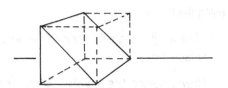

The wedge shape is an example of subtracting from the box shape.

The adjacent drawing started with a transparent one-point perspective box shape. Create the wedge as follows:

❶ Draw a diagonal line from the front top back corner to the front bottom corner.

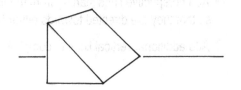

❷ Repeat in the back of the box.

❸ Erase dashed lines and the wedge shape remains.

THE BOX LID

Adding an open lid to the box shape has several extra steps to create an alternate vanishing point (AVP) to form the angled lid surface. You will see this technique used with exterior roof pitches and interior vaulted ceilings. The drawings below start with a two-perspective box shape.

❶ Start with the top front left point on the box and draw a perspective guideline at the angle of the open top. Where this line and the perspective guideline intersect will become the alternate vanishing point (AVP)

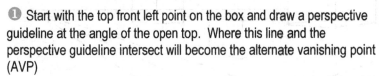

❷ From the top far right corner of the box, extend this line until it intersects with the perspective guideline.

From the left vanishing point, draw a guideline to this intersection.

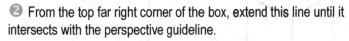

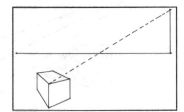

❸ From the top back left point on the box, draw a perspective guideline to the (AVP).

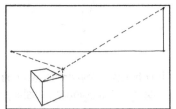

From the left vanishing point (VPL) draw a perspective guideline to through the perspective guidelines from the (AVP). Where these lines intersect will create the lid of the box. Finished box top has the double lines showing the thickness of the box top and the edge of the box. The dotted lines represent the hidden portion of box.

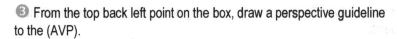

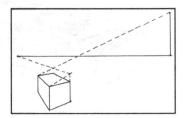

ELEVATION & PARALINE DRAWINGS

Interior design drawings typically show spaces and buildings viewed from several different viewpoints.

One commonly used view is the *elevation view,* which is a side view of an object and does not show depth. The item drawn will appear to be flat and is to scale.

Here two drawings of a box using an *elevation view* of the side and front. Notice the details of each drawing.

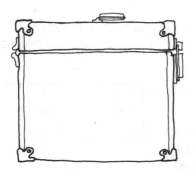
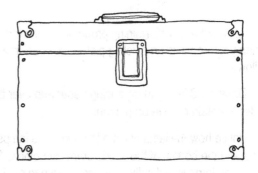

The *paraline* drawing shows an object in three dimension. The *paraline* drawing differs from the *perspective* drawing in that it is to scale and parallel lines do not converge at a vanishing point. There are different types of paraline drawings, which we will not cover in this book. For reference, the one below, is drawn with a 30-degree angle and is called an *isometric paraline* drawing. These drawings are often used in the field when measurements of the components are important, such as when designing open office system furniture.

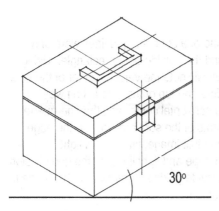

30°

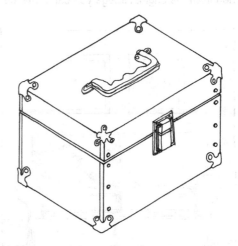

PROPORTION

Now that you are more familiar with different points of view, we are going to introduce the concept of proportion as a technique for creating drawings that are more realistic.

Proportion involves the comparative relationship of size and position between objects or between parts within an object. Good proportion adds harmony and symmetry among the parts of a drawing as a whole. When an object's components are drawn without the correct size relationship, it is out of proportion.

GRID TECHNIQUE

One technique for determining proportion is to start with a grid divided into equal parts. This works well when you are using an existing picture as your drawing inspiration. The lamp in this example was drawn as follows:

❶ Draw a 1/8 inch grid on tracing paper and over the original image in order to obtain relative proportion.

❷ Notice how the lamp fits into the grid. The lampshade is four blocks wide and the bottom of the shade starts five blocks from the bottom of the grid. The lamp is vertically centered relative to each edge. The base of the lamp is three blocks high at its tallest point and the neck of the lamp is about one block tall.

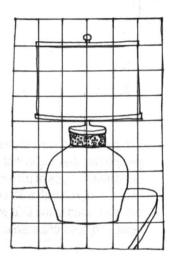

❸ Create a grid with light pencil lines and using a scale that is appropriate for your drawing. The grid may be smaller or larger in scale, depending on the size of the original image and the size of your drawing. In this example, a 1/4 inch grid was created to help create a drawing that is twice the size of the lamp image. Using the proportion information you gained from the original image you can draw the object on your grid.

In this drawing of a sofa, the grid lines were only placed around the outside of the rectangle. Using your eye, you will note the vertical center of the sofa is in the middle, the top of the sofa is half way between the horizontal center and the one-quarter mark. The arms of the sofa are close to the edge. When drawing this image, start with a light rectangular shape and light lines for the grid division. These will provide guides for redrawing the image in proper proportion.

DIVIDING-THE-SQUARE TECHNIQUE

Dividing-the-square is a technique for finding the exact center of a square and thus helping to properly position parts within the box. Here are the steps to use:

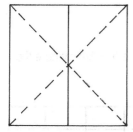 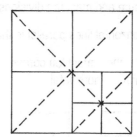 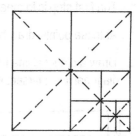

① Starting with the outline of your box, draw diagonal dashed lines from corner to corner creating an "X". This will give you the middle of the square.

② Then draw a cross through the center. The square is now divided into eight triangles.

③ Now continue dividing in the same manner. You will end up with a wall of equal parts with points equidistance apart to act as guides for creating properly proportioned and positioned parts.

The drawing of the box of noodles was created using this technique. Below is a series of drawings showing the progression. After creating guidelines using this technique, the lines and points were used to properly position and proportion the parts.

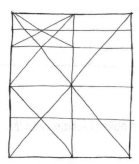 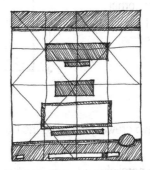 

Helpful Hint:

A technique for insuring the correct proportion and size of a pattern is to use a grid or other guide lines. The measuring tape drawing used light pencil guide lines and rectangular shapes to outline the pattern on the label. This gives you the opportunity to visually check your drawing before moving forward with your marker.

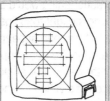 

DIVIDING TECHNIQUES

The drawing steps below provide another technique for dividing a rectangular shape. This technique can be used to proportion the shapes as they recede in perspective.

❶ The first step is to measure and mark the divisions on the left vertical line.

❷ At these points, draw horizontal lines parallel to the top and bottom lines.

❸ Draw a diagonal line from the bottom left corner to the top right corner. Draw vertical lines where the diagonal line intersect with the horizontal,

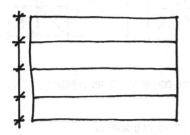 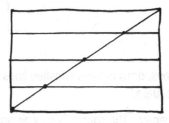 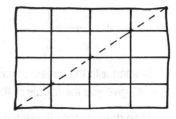

ADDING TECHNIQUES

The steps for extending a rectangular box:

❶ Find the center using the "X" method.

❷ Next, draw a guideline from the bottom corner through the midpoint to meet the extended top line. These two rectangular shapes will be the same length.

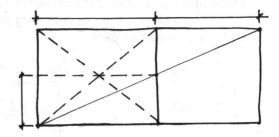

FORESHORTENING PERSPECTIVE

Foreshortening is a technique in perspective drawing to create the illusion of an object receding into the distance or background.

There is an illusion of depth created when parallel lines on a flat surface get smaller as they move away from the front.

A familiar example of foreshortening would be when you look down a long straight road lined with trees, the two edges of the road appear to move towards each other, and the trees appear smaller the further away they are. The adjacent image uses the techniques of dividing-the-box to create a foreshortening perspective.

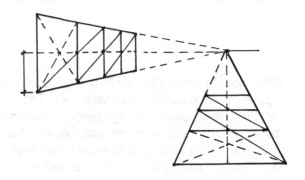

CREATING OBJECTS USING FORESHORTENING

Creating the illusion of depth in a
perspective drawing begins with the
techniques of dividing and adding as
illustrated on the previous page. With a
series of diagonal lines that are created with
the division formula, you can show shapes
that are the same size but appear smaller as
they move from the foreground.

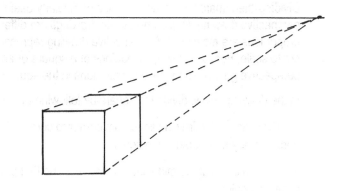

In this series of drawings, a typical one-point
perspective box is the starting point. Follow
these steps to create a series of equal size
chairs displayed in a foreshortened perspective.

❶ Find the center of the side box shape using
the "X" method.

❷ Next, draw a guideline from the bottom
corner through the midpoint to meet the extended
top line. Where these two lines intersect, draw a
vertical line. This creates the correct receding
proportion of that shape.

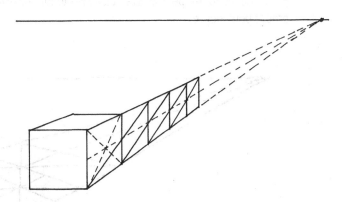

❸ Continue in the same fashion to create
additional receding shapes.

The last drawing demonstrates how to draw multiple chairs in foreshortened perspective using this method.

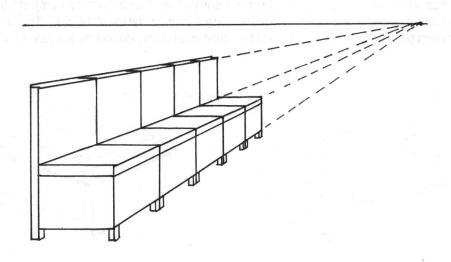

PROPORTION IN PERSPECTIVE

Dividing-the-square technique becomes particularly useful when trying to properly locate points within a perspective drawing. Remember, scale drawings are different from perspective drawings because the proportions are measured. A perspective drawing represents how our eyes see the world naturally, which is not to scale. For this reason, proportions of a square or an object in perspective will be found using perspective guidelines to make the divisions in the square or object.

In the drawing below, divide the square by following these steps:

❶ Draw an "X" by drawing lines from corner to corner. This establishes the middle of the square in perspective just as it did in the square.

❷ Draw a cross at the midpoints by finding a line that projects through the center and toward the opposite vanishing point.

❸ Continue to use perspective guidelines to further divide the square into equal parts, which are also in proper perspective.

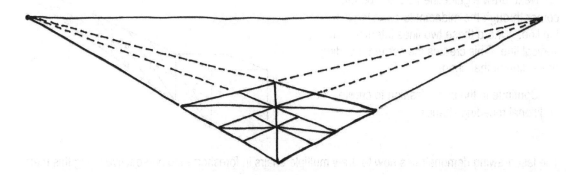

When drawing an object in two-point perspective it is important to have realistically proportioned details. In the example below, perspective guidelines were used to divide the top of the box.   This located the center which provided the correct location of the round pearl on the top and the key hole location on the side.

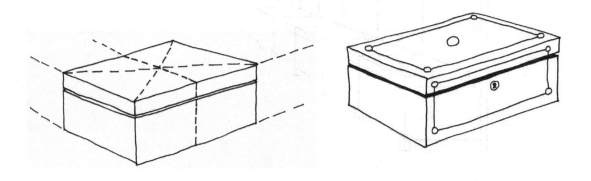

CREATING PROPORTION IN PERSPECTIVE

Here is an exampe of using the "X" technique to illustrate ribbon on a box.

① Start with the top of the box and locate the center by using the "X" technique.

② Draw the edges of the ribbon in using the two vanishing points.

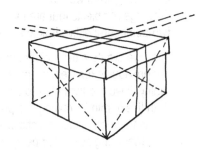

③ Where the ribbon edges meet the side of the box, add vertical lines to create the ribbon on the sides of the box.

④ Draw another line parallel to the ribbon edge to create depth.

⑤ Two free-form loops create the bow shape, with the center knot and the two ribbon lengths draping over the box.

⑥ In the finished drawing, vertical and horizontal lines were added to create contrasting values between ribbon and box.

Here are several objects shown in two-point perspective. Notice the details added to each object continued to use the vanishing points for locating the vertical lines. The extra perspective lines are not shown in this image and yet they were used to help create proper proportion and perspective..

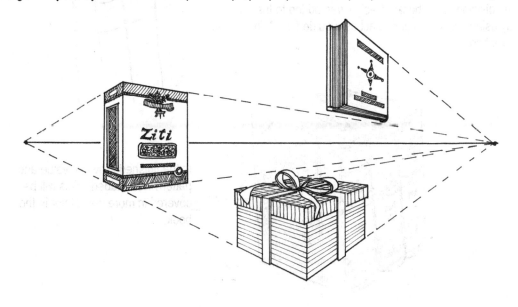

ADDING DETAIL TO THE BOX

Combining what we have learned about drawing in perspective and proportion, we can now create detail drawings of objects that are more elaborate. The drawing series below shows how to move from a simple two-point perspective box to a finished drawing of three books.

First, we create a two-point perspective box, but this time using vanishing points that are more realistic. In our previous exercises, we created boxes with vanishing points only inches away from each other. In reality, if we are trying to represent a realistic perspective, the vanishing points wound be several feet apart. With this in mind, drawing two-point perspective objects like books, requires using "imaginary" vanishing points to create more realistic images. You will need to start imagining the vanishing points in order to obtain the proper angles of your perspective lines.

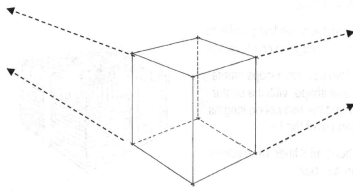

The image was started by drawing the leading edge of the box. The top and bottom perspective lines were drawn using "imaginary" vanishing points that are off the page.

The arrows show the angles of the imaginary perspective guidelines.

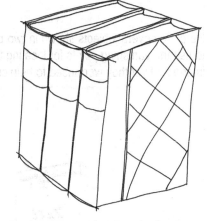

The box was then divided into three books using the "imaginary" vanishing points. The left side plane was further divided to depict the binding design. Additional lines were added to the top plane to show the division of the books. Design was added to the right plane by using a diagonal grid to act as a guide for adding pattern for the front cover.

In the finished drawing, value and pattern were added. This will be covered in more detail later in the book.

SKETCH PRACTICE PAGE

Using your sketchbook page as a place to play and practice allows you space to draw different objects more freely.

Try coming up with an unusual point of view for your sketch. Consider a close-up view or views from above or below.

Having fun with your sketch time will keep your technique loose.

PUTTING IT TOGETHER

It is exciting to start using your hand drawing skills to generate ideas from sketches. Using a single subject for practice is a good technique to help develop ideas. Here is a series of images that show the process of sketching to a finished drawing.

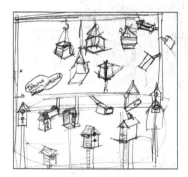

In this sketchbook page, I was exploring ideas for a box shaped object that I could use in a drawing. This included hanging planter boxes, light fixtures and birdhouses. This type of sketching is a method for thinking visually. Instead of talking through your design ideas, you are drawing shapes and objects, quickly filling up the sketch page.

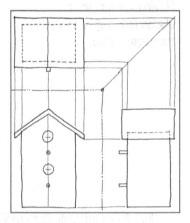

After playing around with these images, the birdhouse image became the focus for the final drawing. To visualize more clearly the proportion and parts of this shape, a multi-view image was drawn.

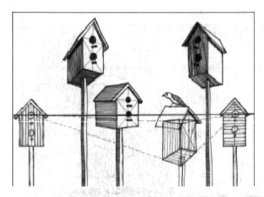

It actually took two drawings to get an image that was successful. After evaluating this first drawing, several elements were unsuccessful. The birdhouse with the bird perched on top was too confusing to understand.

Here in the final drawing, I added more interest to the birdhouses on the side and the transparent birdhouse is reading more successfully. Leaf shapes were included to add interest.

Creativity Strategy
LOSE THE REFEREE

When you are starting to learn how to draw, it is easy to get stuck by your own expectations. In the beginning, your images may not look like the object, they may be out of proportion, and your drawings may not be as well as another student. If you are just leaning to draw, it is unreasonable to expect your drawings to be perfect. It is unreasonable for you to expect to be as good as someone who already has some drawing experience. Unreasonable expectations can become a huge obstacle to learning, especially if they are preventing you from moving forward. If you find yourself being too critical, try to image that you are in a kindergarten class learning to draw. When we were children, we had little expectation and therefore we were not afraid to explore. Drawing was fun because we were all artists.

Consider what happens in a playground. The schoolroom door opens and out rush 15 kindergarten students who have been inside sitting at their desks all morning. They scatter through the playground - a couple of students on the swings, a few in the sand box and some going down the slide. Next thing you know, they are spontaneously jumping over to a different area. Are the students learning as they play? Of course they are. Is anyone scoring their play? Does the teacher have a clipboard to note progress on their play with a grade? Is there a referee blowing a whistle? No! This is time to explore, experiment, and enjoy trying new activities.

Allow yourself the freedom to play around with your new drawing skills. Loose the referee as you practice making a line, drawing boxes and adding texture. This attitude of being open and allowing yourself to explore new activities will serve you well as you take on the challenge of learning to draw. Have fun and go play!

Chapter 3

Cylinders

Now that you understand how to draw the box shape in different perspectives and correct proportions, we will learn how to subtract portions of that shape to create cylindrical containers.

A cylinder is simply a form based on a circle and drawn in perspective.

GETTING STARTED

It is important to be able to understand how we see cylinder shaped objects in order to create realistic perspective drawings of them. We begin by learning how a circular object changes in appearance as we change our viewpoint. Then, we will learn how to apply different techniques for creating cylindrical objects in perspective. Once you have mastered the techniques you will be able to draw cylindrical objects more freely and with fewer guidelines.

Key concepts for this chapter include the following:

Circle. A two-dimensional geometric round shape.

Ellipse. A circle drawn in perspective.

Cylinder. Object or shape with circular ends. A tube-like object.

Contour lines. Lines that precisely following the curves and planes of the object.

Major axis. Line extending through the widest point of an ellipse.

Minor axis. Line that appears to extend through narrowest points of an ellipse and at a 90-degree angle to the major axis.

Center line. The line that extends through the center of a cylinder.

CYLINDER DRAWING TECHNIQUES PREVIEW

There are two common techniques for drawing cylinders - one begins with the box, the other with the centerline or minor axis.

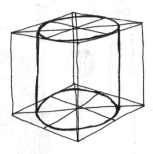

Divide-the-Box Technique. In this technique, we will use what we have learned before about creating boxes and proportion and apply the same concepts to creating circles and cylinders in perspective.

Center-line Technique. The second technique uses a method of drawing ellipses with a major and minor axis about a centerline to create cylindrical shaped objects. Understanding the relationship between major and minor axis and the centerline helps us more easily create freehand drawings of perspective cylindrical objects.

CIRCLES & ELLIPSES

A circle is a geometrically symmetrical round shape. Circular shaped items can also be viewed as elliptical shapes. A circular object, like a penny viewed on one side, will become an ellipse as it is rotated. In addition, the circle appears elongated when it is viewed as an ellipse.

The illustrated pennies show how the shape will shift as the viewpoint of the object shifts. The penny at the top appears as a circle when viewed flatly and at eye level. As the penny is turned, it becomes an elliptical shape. It appears flat when viewed from the side. As you move through this chapter, keep this concept in mind. Take time to look around at cylinder shaped objects in your environment and notice how the elliptical shapes change as you change your viewpoint.

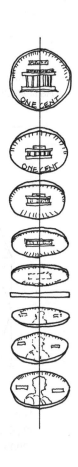

BASIC DRAWING TECHNIQUE

One of the fundamental approaches to drawing circles is to start with a square box shape. Guidelines are drawn to provide points of reference. These guidelines are created by dividing the square from corner to corner with the "X" and then drawing a cross through the center. Next, the "X" guidelines are divided into thirds and the outside third is marked. Notice in the second drawing how the outside guide points on the "X" and the outside points of the cross will become the guide points for drawing the circle.

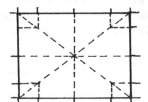 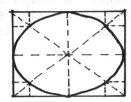 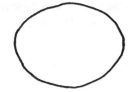

The same approach is used to draw an ellipse. In the drawing below, a rectangular shape is drawn first to form the basis of the ellipse. The guidelines and points are created using the same technique as above.

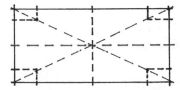 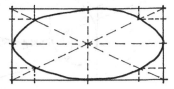 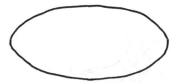

Helpful Hint:

When drawing circles, ellipses and cylinders it is best to use a soft lead pencil, such as a 2B pencil. Softer lead will move easily over your paper assisting in making the curved shape. Your drawing markers are also a good drawing tool to use to draw circles. You can start your drawing with pencil guidelines, finish the cylinder shape using your marker and then erase the pencil guidelines.

ELLIPSE IN PERSPECTIVE

Fundamental to understanding elliptical shapes is seeing how their shape changes in a perspective view. As illustrated by the image below, an elliptical shape is determined by its location in a perspective view.

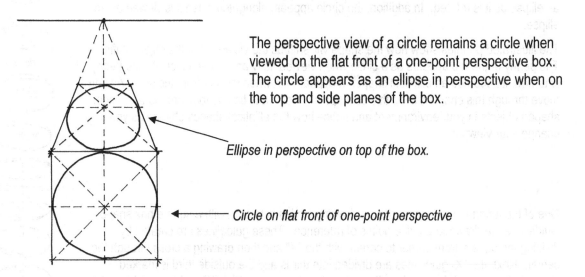

The perspective view of a circle remains a circle when viewed on the flat front of a one-point perspective box. The circle appears as an ellipse in perspective when on the top and side planes of the box.

Ellipse in perspective on top of the box.

Circle on flat front of one-point perspective

Understanding the relationship of the major and minor axis of an ellipse is part of correctly drawing an ellipse in perspective. The major axis is an imaginary line found at the widest part of an ellipse and the minor axis is found at the narrow diameter of the ellipse. The minor axis appears at a right angle to the major axes.

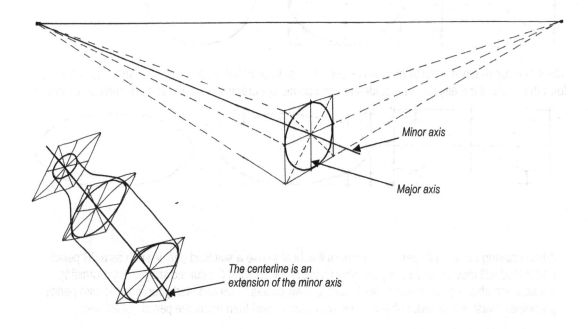

Minor axis

Major axis

The centerline is an extension of the minor axis

DRAWING CIRCLES & ELLIPSES

The detailed steps for drawing a circle or an ellipse are basically the same. The only difference lies in the starting shape of the box. To draw either a circle, a one-point perspective ellipse or two-point perspective ellipse, follow these steps:

❶ Start by drawing a square, a one-point perspective box, or a two-point perspective box.

❷ Draw an "X" from corner to corner.

❸ Draw a cross from the center of the outside edges and through the middle.

❹ Create guide points by splitting each arm of the X into thirds and marking the outer third.

❺ Mark additional guide points at the end of each cross where it touches the box edge.

❻ Draw a circle by creating curved lines from guide point to guide point.

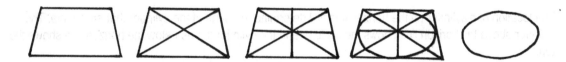

Guide points

To create a one-point perspective ellipse, put the circle in a one-point perspective square.

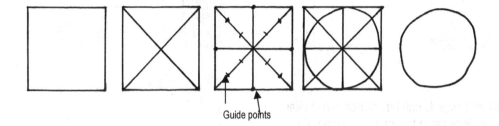

To create a two-point perspective ellipse, put the circle in a two-point perspective square.

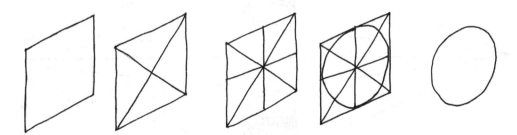

DRAWING CYLINDER SHAPES

The concept of drawing cylinders can be applied to other cylinder shapes. Here are demonstrations of how to use the two-point perspective box as the starting point for drawing these shapes.

❶ In light pencil, draw a two-point perspective box the length of your object.

❷ Draw the elliptical shapes on the top and bottom using the divide-the-box technique.

❸ Connect the guide points of the ellipses with vertical lines to form the sides of the object.

❹ Draw the contour lines of the object with your drawing markers and the erase pencil lines.

❺ Add a second ellipse shape to form the rim and additional parallel lines to finish the drawing.

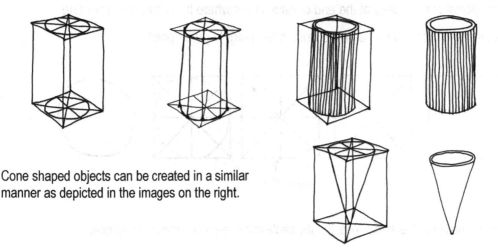

Cone shaped objects can be created in a similar manner as depicted in the images on the right.

The location of a cylinder on the horizon line will determine the parts of the cylinder that are visible. The cylinder above the horizon line shows the bottom, and the cylinder shape below the horizon line shows the top.

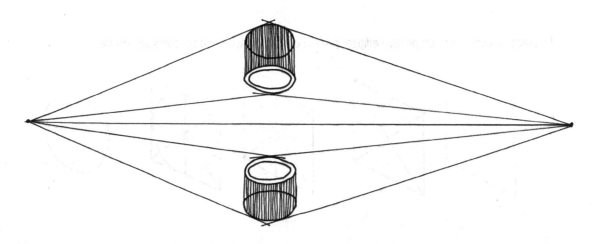

CYLINDERS WITH DIVIDE-THE-BOX TECHNIQUE

Here are steps to take when drawing cylinders starting with a two-point perspective box. You will be using this concept throughout the chapter. Take time to learn the following steps.

❶ In pencil, draw a rectangular shape box the size of the cylinder.

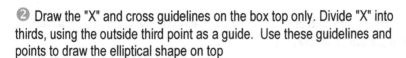

❷ Draw the "X" and cross guidelines on the box top only. Divide "X" into thirds, using the outside third point as a guide. Use these guidelines and points to draw the elliptical shape on top

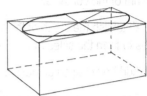

❸ Draw a vertical line on the "X" and cross guidelines where they touch the elliptical shape down to the base of the box shape. Use these guide points to divide the "X" into thirds, using the outside third point as a guide

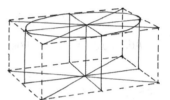

❹ Connect these points to draw the elliptical shape on the bottom.

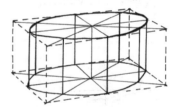

❺ Erase the back of the container guidelines and add the lid lines.

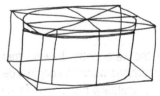

❻ Use your drawing marker to outline the contour or outside shape of the container.

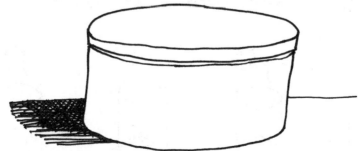

MULTI-DIAMETER CYLINDERS

Some cylinder shaped objects have two different diameters. For example, the flowerpot consists of two separate cylinder shapes - the top portion has a larger diameter shape and the bottom portion has a diameter that tapers from top to bottom. The drawing below demonstrates how guidelines can assist with drawing the flowerpot in perspective.

① Start by drawing a rectangular shape and then use the divide-the-square technique to establish guide points for creating the elliptical shape. Use these guide points to draw the elliptical shape that will form the top rim of the container.

② Establish the depth of the top portion and then draw a curved line parallel to the top curve. Add vertical lines to form the sides of the upper portion of the pot.

③ Add box shaped guidelines to establish the base shape.

④ Draw curved lines from guide point to guide point defining the base shape.

⑤ In marker, draw the contour or outline of the pot.

⑥ Add a plant shape and pebbles to finish the image.

MULTI-ELLIPSE CYLINDERS

In the water bottle drawing, there are three different elliptical shapes. These shapes were drawn from guidelines within the original perspective box. These elliptical shapes were connected with vertical contour lines imitating the shape of the water bottle. Parallel lines were used to add the shape of the label and to add detail to the top.

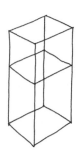

COMPLEX CYLINDER OBJECTS

A mug is a more complex cylinder shaped object that has a box shaped handle on the side. When drawing a mug start with the cylinder shape. Start with the same steps as used for drawing the flowerpot container.

❶ Draw the square shape at the top of the mug. Divide with the "X" and the cross to provide guide points for assisting with the ellipse shape.

❷ Draw two ellipses, one within the other, to represent the rim of the mug.

❸ Draw a parallel curved shape for the bottom of the mug.

❹ Add lines for the side of the container.

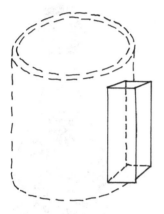

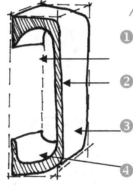

Adding the Handle

❶ Draw the rectangular shape on the side of the mug.

❷ Draw the shaded shape on the front of the rectangle as a guide.

❸ Draw the outside plane using the top and side of the rectangle as a guide.

❹ Draw the inside planes under the top and on the bottom of the handle.

The contour drawing shows the completed mug and the last drawing illustrates the addition of value and a surface.

CYLINDERS WITH CENTER-LINE TECHNIQUE

Another technique for drawing a cylinder begins with drawing the centerline of the object. As noted in the beginning of the chapter, the centerline or minor axis is an important aspect of the ellipse and cylinder. To draw cylindrical objects using the centerline technique follow these steps:

1 In pencil, start by drawing a centerline longer than the length of the object.

2 Add light horizontal pencil guidelines at points where the object changes diameter and drawn with scaled distance from the centerline.

3 Add light pencil contour guidelines of the object by connecting the horizontal lines.

4 Draw darker pencil ellipse lines creating the curved shape.

5 Redraw darker contour lines of the object from the edges of the ellipses.

6 Add marker to finish the drawing and then erase any remaining pencil lines.

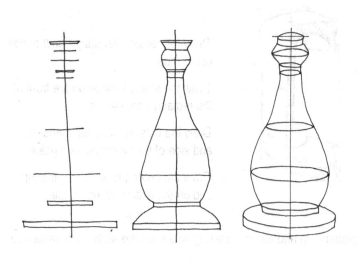

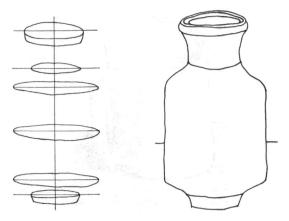

Helpful Hint:

When drawing the contour lines on the outside of the object shape, remember to move your eye ahead from one line to the next. This is similar to what you practiced in Chapter 1. As you practice moving your eye ahead of your pencil or marker, you will improve your sketching skills.

DRAWING LAMPS WITH CENTER-LINE TECHNIQUE

In this series, the centerline is the starting point for drawing the lamp.

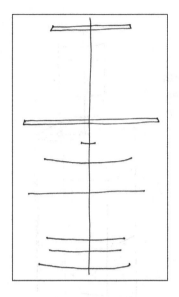 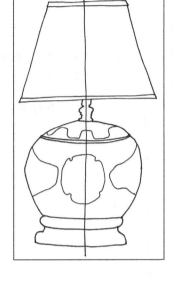 

❶ Centerline down the middle with horizontal lines placed proportionately down the lamp, provide guidelines for both the width and height of each section.

❷ Contour lines define the overall shape of the lamp and the outline of the pattern on the base

❸ The final marker drawing has value and pattern added to the base of the lamp. Add more interest to the image by adding items around the lamp and background.

A centerline assists with the proportions when drawing this statue. It provided a guide for the length of the arms on each side, the face, robe, feet and base. These beginning guidelines were in light pencil. Marker was added on top of the pencil lines.

COMBINING TECHNIQUES

In this drawing series of a water cooler, a two-point perspective box was the starting point for drawing the base shape and a centerline provided guidance for the water container on top.

❶ Leading edge box with the top divided and a centerline drawn with pencil.

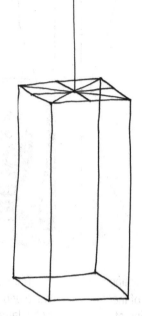

❷ Center line with circular top. Lower box shape divided into three areas.

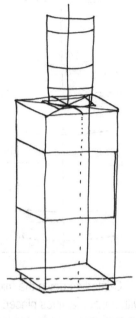

❸ Marker used to draw outline shapes and details.

❹ In the final marker drawing value is added by including stippling on the water cooler and adding a background and flooring.

Chapter 3 - Cylinders

SKETCHING PRACTICE PAGE

Taking time to sketch objects with cylinder shapes will be part of the success that you will have in your drawings. Sketching gives you an opportunity to coordinate your hand and your mind together to draw these shapes. These two drawings were practice pages that were done spontaneously just for fun.

It can be beneficial to use a picture of the object that you are drawing. You can grid the outside of the picture to assist with the proportions of the image. Here is a series of images that includes the starting picture, a quick sketch and the finished image.

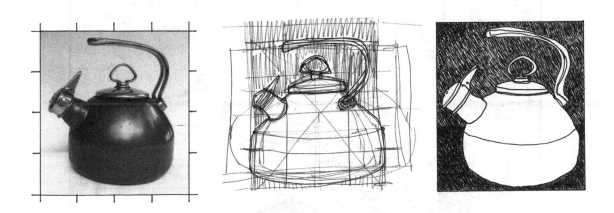

PUTTING IT TOGETHER

Drawing cylinder shapes has always been a challenge for me. In this drawing from my sketchbook, I am playing around practicing and doodling with cylinder shaped objects. It was easy to sketch objects that were close to my work area and I sketched objects all over the page. The hardest part of drawing realistic cylinder objects for me is getting the top and base ellipse shape to look correct. I often need to draw guidelines to get the shape and proportion to keep the object from looking discombobulated.

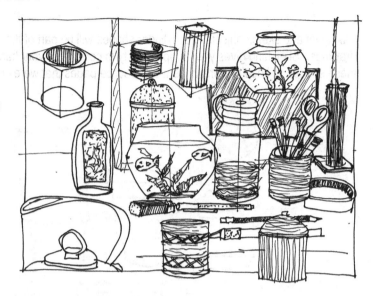

Often this type of free sketching practice will generate ideas for a finished drawing. From the sketch above, I used some of the items and created the finished drawing below.

Creativity Strategy

STRIVING FOR PERFECTION

A perfectionist is a person who "strives for or demands the highest standards of excellence in work, especially in his or her own work."

Striving for perfection can be both good and bad. The perfectionist drive is beneficial when it pushes you to do better. On the contrary, the perfectionist drive can also paralyze the creative process.

The creative process necessitates a willingness to be open to new and different outcomes, not simply one idea of "perfection." Therefore, the goal of any drawing should be to do the best you can at that moment and then move on to the next drawing. Your drawings may never be "perfect," but they can be successful if you set realistic goals.

When striving for perfection, look at your goals in perspective. Identify who is deciding perfection, consider your current level of skill, and make sure your idea of perfection is reasonable.

A perfectionist voice can help you improve if it says: "I want to try this again, I think it will be more successful the second time". However, a perfectionist voice that says: "This is not perfect, it's not good," can paralyze your progress. Strive to improve your skills, but also be able to put aside a perfectionist voice that prevents you from being open to different outcomes, and thus can be discouraging as you practice.

One way you can push aside the perfectionist voice is to allow your drawings to be finished when you think you are 80 percent done. This way, you will learn to accept varying outcomes, and then be able to move on to new drawings.

Drawing Notes

Keep in mind, freehand drawings are not the same as computer generated or photographic images. People are often attracted to freehand drawings because they are "not quite right." Being drawn by hand gives them more character than photographic perfection. For example, see the camera drawing on this page. The drawing is not perfect. The lens looks like it is melting, and yet this makes the image more interesting than one that is photographically perfect.

Chapter Four

Texture & Pattern

Now that you are gaining more confidence and skill, you can begin to add value to create texture and pattern in your drawings. Adding value will define the form of the shape to the contour line drawing. It also provides an opportunity to add your own artistic character to your pieces. Now, we will use line to add material qualities of surface and volume. This will involve using tonal values. These values will convey a sense of light, mass, shape and texture. Within this chapter, you will continue to use pencil and drawing markers, but these skills will be easy to transfer to other mediums when you are ready.

GETTING STARTED

In this chapter, we will focus on value as it used to define actual texture, implied texture, and pattern. How we see or view an object is determined by the patterns and tones created by light. Value helps identify the shape and form of objects by suggesting varying tones of light and dark.

The following concepts will help you throughout Chapter 4:

Value. The range of possible lightness or darkness achieved with a marker or pencil. One of the first steps to adding value to the drawings is to determine the light and dark areas.

Range. The extremes from black to white and shades of grey in a tonal scale applied to the drawing.

Relativity. The degree of comparison of one to another as applied to darks and lights of the object.

Contrast. The result of comparing one thing to another and seeing the difference.

Line weight. The thickness of a line.

Texture. The surface quality of the object.

Pattern. The arrangement of forms or shapes that create an orderly whole.

Abstract Pattern. Pattern that is altered, simplified or changed and yet is still recognizable.

Implied Pattern. Pattern that is suggested or accented to create interest in the drawing.

Rendering. Artistic portrayal of and object and visual communication of form through value changes.

Rendering Pattern

Rendering can be used to show an object's pattern. The grinder's metal's finish has used stippling as the texture.

Implied Pattern

The graphic design on the cream container is rendered with lines and dots.

Rendering Texture

Rendering can be used to show texture. The texture rendered to the mug was the actual pattern

LINE & RENDERING

While there are many different techniques for drawing, *line* drawing utilizes an architectural style for rendering images.

Rendering is a common term in architecture and interior design to describe the visual enhancement of drawings. It is used to better communicate the subject visually and can assist in making the subject easier for the viewer to understand. Rendering adds visual strength and interest by creating a three-dimensional quality to a drawing.

Architectural style of rendering incorporates specific techniques of utilizing lines and dots to create the illusion of texture, pattern, and dimension.

The adjacent table demonstrates a number of examples of this technique.

Helpful Hint:

As you add rendering to your drawings, keep in mind the importance of contrast in creating interest.

Our eye is naturally attracted to contrasting values. It will seek out the very dark and the very light parts in the drawing.

Adding contour lines to define the outside edge of your subject is another technique for adding "pop" to your drawing.

Contour Lines			
Parallel Lines			
Cross-hatching			
Dots or Stippling			
Curved lines			
Wavy lines			
Broken lines			

VALUE

Value refers to the range of possible lightness or darkness achieved with a marker or pencil. It is often expressed in gradations moving in degrees from light to dark. Common terms used to describe these gradations: white or highlight, light gray, medium gray, dark gray and very dark gray or black.

In the value scale below, the first block is the highlight, which is white. The next four blocks gradually move from medium dark to very dark. When you are first starting to add gradations of value, keep it simple and limit yourself to three degrees of grey tone plus a white and a very dark or black.

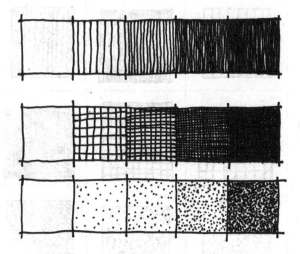

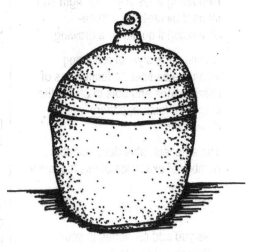

Also, as you add value to your drawings, vary your line weight. Use dark or thick lines to define the contour or outline of your object and thinner, lighter lines when adding value, texture or pattern.

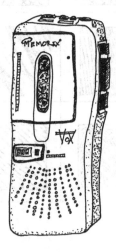

Here are examples of one-point perspective objects using a variety of value and pattern to illustrate their features.

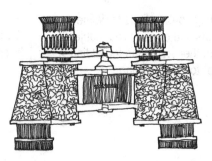

Chapter 4 – Texture and Pattern

CONTRAST

The design principle of *contrast* is defined as the result of comparing one thing to another and seeing the difference. You can create it in a line drawing by using a variety of patterns textures and values. The two drawings below demonstrate how this principle is used to highlight the design element of the image. In each image, contrast exists with the difference in the value of the background and the objects.

In the first drawing, the dark value of the background is created with vertical lines on the wall and wood texture pattern. The foreground objects are in a contrasting lighter value. The table is white; the lamp has a white shade and light diagonal lines. The basket has a crisscross pattern and the artwork has light stippling texture.

In the second drawing, the value contrast alternates. The background is white and wood pattern has a light value. The table and objects have a darker value. The patterns and textures include a dark crisscross on the lamp base, wood lines on the table and a dark value on the picture frames.

TEXTURE

One of the ways to add value is to incorporate texture in your drawings. Texture refers to the surface quality of the object you are drawing. It creates a visual feeling that you may describe as smooth or rough. It is a design element representing a portion of the overall creative solution to a drawing.

You know the texture of an object by how it feels to the hand. With your drawings, you will be exploring ways to use line, stippling and patterns to depict visually the texture of your object. The appearance of texture is created using techniques outlined earlier in the chapter. Lines can vary from being short, long, jagged or curved. Stippling is created by using multiple dots in groupings. Circles can also be used to define a textured object.

There are two types of texture - actual and implied. Actual texture attempts to mimic real qualities. Implied texture utilizes stippling or line to suggest texture on the object.

IMPLIED TEXTURE

An object may not have an obvious texture to simulate in your drawing. In this case, lines or dots combine to show form, texture and material. These are often called modeling lines. In the drawing, the contour lines show the shape of the object and the texture lines or dots create a visual pattern. Notice implied texture with modeling lines in the examples below.

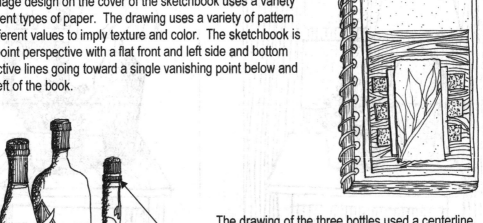

The collage design on the cover of the sketchbook uses a variety of different types of paper. The drawing uses a variety of pattern with different values to imply texture and color. The sketchbook is a one-point perspective with a flat front and left side and bottom perspective lines going toward a single vanishing point below and to the left of the book.

The drawing of the three bottles used a centerline technique to draw the beginning shapes. Each part of the bottle and of the ground uses a different value technique. Notice the following aspects:

1. Parallel lines used to add value to the bottle tops and labels.

2. Cross-hatching used to show a shaded value.

3. Stippling used to form the ground surface.

ACTUAL TEXTURE

In the example below, value helps depict the visual qualities of surface and volume. Notice how the last drawing uses value to convey a sense of light, mass and space.

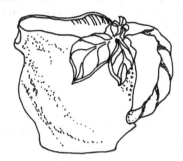

In the first drawing, the pitcher appears flat with only the contour line. The second drawing has texture added and the pitcher begins to take on a three-dimensional shape. The change in value is still very slight which keeps it from being dynamic. The third drawing is more interesting and easily recognizable as a pitcher. It is completely rendered with a variety of values.

WOOD SURFACE TEXTURE

An important aspect of rendering wood patterns is to identify the features and variety of values in the wood. The three examples on the right demonstrate how different types of wood have visual features that can be rendered. The variety of pattern is created by the varying line weights, the curvilinear properties of the line, and the contrast of value. These are three plastic laminate samples - the first one is river cherry, the second is empire mahogany and the last one is limber maple

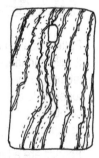

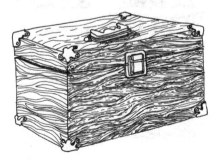

The box illustration with rendering that depicts a wood texture. There is a single source light coming from the top left side effecting the value or shade of the box.

The rendered textures shift slightly on each side to depict the different light vales. The front of the box has a darker value and more detail rendering with lines and stippling. The left side shows only the patterned lines and it is slightly lighter. The top suggests the patterned lines and has the lightest value.

PATTERN

The design principle of pattern is defined as the arrangement of forms or shapes that create an orderly whole. In your drawings, using pattern provides an opportunity to add value, line and shape. There are many different types of patterns including fabric patterns, decorative patterns and patterns of natural material. It is important to draw patterns with the correct proportion and size relative to the size of your object or surface.

PATTERN TYPES

Motif. A repetitive decorative design, shape, or pattern.

All-over pattern. Using of a motif in a planned or predictable way.

Random pattern. Using a motif throughout the object in an unpredictable, scattered pattern.

Pattern is an important element in the design of interior spaces. It will coordinate color schemes, and add character, scale and interest to the space. Multiple patterns can be combined together to create interesting and exciting interiors. This image combines different motifs unified with the blue color, white background and curvilinear patterns

Fabrics are a typical way to add pattern into interior designs. In your drawings, you can render actual pattern using straight, angular, or curved lines. The appearance of the fabric patterns will depend on the type of line, shape and values that are used.

Fabric patterns will often have a number of motifs, or single-design units, arranged into a larger design composition. Patterns may be read as texture if they are very small.

Techniques for drawing fabric patterns on furniture will be further discussed in Chapter 6 - Furniture

Chapter 4 – Texture and Pattern

COMBINING TEXTURE & PATTERN

There is a combination of texture and pattern in this drawing, which adds interest to the image. The variety of line, stippling and geometric shapes shows each item uniquely. Notice the following details:

① The milk carton image has vertical lines with different values along with a drawing of the barn shape.

② The grinder image is rendered with stippling to imply a shiny surface.

③ The ceramic mug is rendered with a textured geometric pattern that was actually on the mug surface.

④ The spoon is rendered with parallel lines of varying values.

Here is another example of combining various textures and patterns to render the different surfaces.

① The lamp has a motif pattern.

② Wood rendering is applied to the picture frames and tabletop.

③ The plant container incorporates a variety of texture and pattern to add interest to the drawing.

SURFACE RENDERING

In an interior design or architectural drawing, it is often important to be able to communicate specific surface patterns and textures as part of the design. The samples below are renderings of surface materials that reflect a combination of actual texture and pattern.

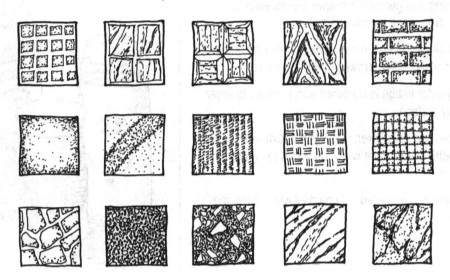

Practice Rendering Sketch

In this quickly sketched rendered floor plan of a bedroom, it has a stippling pattern representing the texture of the carpet. There are lines and dots for wood grain and horizontal squiggled lines representing the fabric on the bed. Around the floor plan, I doodled different tabletops, plants and flooring finishes.

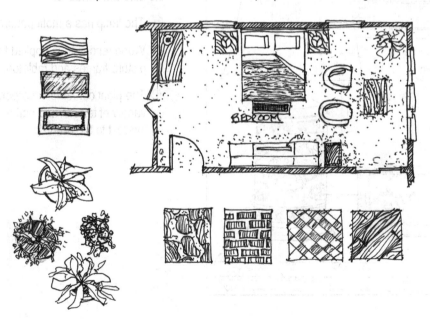

FLOOR PLAN SURFACE RENDERING

The addition of surface rendering to floor plans provides the opportunity to show the pattern and texture of the building elements and furnishings. This is a simple way of successfully distinguishing material qualities in an interior space. Materials are rendered in scale with indications of surface finishes. It is best to carefully examine the actual material and render an impression of what you see. The flooring and furniture finishes are often rendered on the same drawing. On this page, flooring surface rendering has been included. Three main rendered flooring textures include carpet, tile and wood plank.

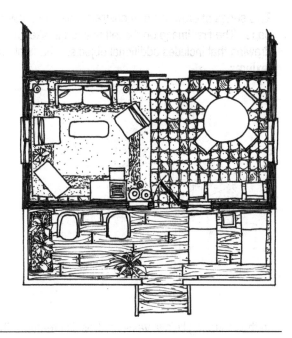

This second floor plan has rendered fabric patterns and rendered wood grain on case pieces. The sun on the top right corner indicates the direction of light that is supported by the darker shade and shadow value added to the objects.

PUTTING IT TOGETHER

This series of drawings demonstrates the steps used to complete the desk drawing on the bottom of the page. The first image on the left is a rough sketch of the drawing idea. The second image is a contour line drawing that includes additional objects. The final drawing is rendered with a variety of value, patterns, and textures.

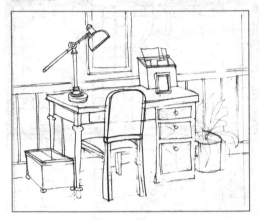

In the rendering below, value contrast is created with the dark paneling against the light desk. This highlights the desk and the plant. A right side light source is suggested by the shadow and shading rendered on the plant container, below the desk, chair and basket. Surface patterns depicted in the drawing are shown in the blocks on the right. They represent surfaces of the wall, the chair, the basket on the desk, the basket on the floor and the paneling.

Chapter 4 – Texture and Pattern

Creativity Strategy
TURNING THE PAGE

When I decided I wanted to learn to draw, I realized I had to take some lessons. I recognized I would benefit from the instructions, the homework assignments and having this activity on my schedule. So, I turned to one of the local art museums that offered drawing courses. I began with Drawing I. At first, I was embarrassed because many of the other students seemed to know what they were doing, and I didn't have experience in drawing art. I pushed myself to keep going to class anyway, even though it was hard being the worst student.

After Drawing I, I should have signed up for Drawing II, and yet it wasn't being offered. So I took Drawing I again, this time with a different teacher.

On the second try, I was much more successful. However, when it came time for me to move on to Drawing II, it was not offered again. However, since I still did not feel very good at drawing and I learn more in a class setting, I signed up for Drawing I for the third time with the original teacher.

"What are you doing here?" the teacher asked on the first day of class. "You already took this class".

I explained that I needed practice and was looking forward to taking a class. She gave me my own assignment: To draw my German shepherd, Sadie, in my sketchbook repeatedly. In addition, this time there was to be no erasing. This sounded fine until I got out my sketchbook and my pencil and looked at the dog.

I panicked. Sadie is big, she moves around and there are so many different parts to sketch. At the library, I found a few books on drawing dogs, and I started on smaller parts. First, the paws, then the foot, then two paws.

I found as I worked on a drawing that I did not think was going in the right direction, I would turn to a new sketchbook page and start again. This lesson became the best tool I have found for dealing with self-criticism and a great metaphor for dealing with drawing or with other life issues. If you don't like what's happening and you can't erase what's been done, just turn the page.

For each Drawing I class I collected my drawings in my sketchbook and kept them all. It was fun to go back, review my work from the beginning, and start to recognize the progress I had made simply by continuing to work on what I had already learned and continuing to turn the page.

Chapter 5

Shade &
Shadow

In this chapter, we will explore another aspect of value by examining how a light source gives shape to objects through the addition of shade and shadow. We will continue to use a variety of line texture and patterns.

GETTING STARTED

As we learned in the previous chapter, light is a major factor in how we see the world as it creates varying degrees of light and dark values. The gradation of values created by light falling on an object creates shade, shadow and shape. The depiction of light in your drawing determines how the image is interpreted. Effective rendering of these gradations of values in your images will provide a more realistic quality to your drawings.

These are key concepts that we will address throughout Chapter 5:

Shade. The darker value on the object itself, which is farther away from the light source.

Shadow. The image on the ground cast from the object, which is a dark value

Shape. The outline of the object, which creates form.

Form. The shape or structure of an object. The illusion of volume or mass in a drawing.

Gradual. To change slowly in steps or degrees.

Ground. Surface that an object is sitting upon.

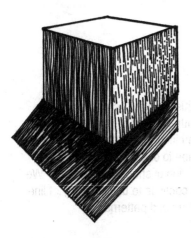

Value can be used to add shade and shadow to an object. In the box drawing on the left, there is a single source of light above and to the right side. This location of the light determines the different values on the box and the shape of the shadow on the ground.

The jar drawing on the right demonstrates how gradual changes in value will create the illusion of three dimensional form and shape. The drawing depicts a single light source from the front by using value changes that gradually get lighter as you move from the sides and toward the center.

RENDERING SHADE AND SHADOW

Shade refers to the value (lightness or darkness) on an object and shadow refers to the value on the ground. In reality, there are infinite degrees of value; however, for simplifying our drawings it is common to reduce these gradations to a limited scale ranging from white to black. For our beginning line drawings, I suggest limiting your value gradations to a range of four or five. In the drawing below is an example of a gradated value scale with four degrees. Using this scale on the box shape below, the top is white (#1), which is closest to the light; the left side is light gray (#2); the right side is medium gray (#3) which is further away from the light; and the shading is dark gray (#4), which receives no light.

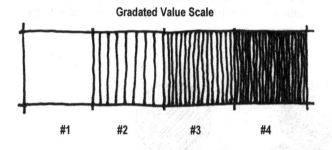

Gradated Value Scale

#1   #2   #3   #4

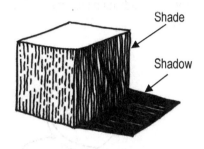

Shade

Shadow

The example on the right shows how to add gradated value to your objects using an architectural style. Rather than being scribbled or quickly added, each line on the drawing is deliberate and touches the edge of the contour lines. The illustration to the right using these steps:

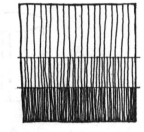

❶ The first set of vertical lines extends from top to bottom touching the contour lines and leaving space in between each line.

❷ The second set of lines start at the guideline and are in between the first set of lines.

❸ The third set of lines also start at the guideline and are in between the second set of lines.

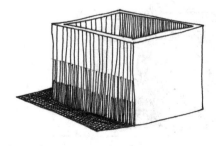

In this example of an open box, the light source is on the top right side. Notice how line is used to depict shade and shadow.

❶ The right side where the light is directly pointing is the lightest value.

❷ There is a shadow on the inside of the box depicted with dense lines.

❸ The left side is the shadow side. It graduates in value, from light on the top, to dark on the bottom.

❹ Shadow is on the ground and located on the right and back sides. It also slighting graduates from dark, close to the box, to lighter, away from the box.

RENDERING CURVED SURFACES

In the shapes drawn below, each rendering of shade and shadow uses different techniques. The addition of shading to these objects provides dimension and shape to the drawing. Shading of curved surfaces is similar to shading a flat surface; however, since the surfaces are not separate, shading of transitions occurs by gradually moving from light to dark.

In the curved shapes below, a four value shading scale helps to create the illusion of a single light source from the left side. Each graduation of grey shows a change in value. The shadow on the ground is darker than the shading on the object. The top row of shapes shows the transitions points from one value to the next.

Helpful Hint:

Cast shadows can also have a gradation of value with the darkest value shown closest to the object. The shadow will graduate from dark to light as it gets further away from the object.

PERSPECTIVE CAST SHADOWS

Adding cast shadows in your drawings will create a more realistic image. There are varieties of techniques to show cast shadows. In the two examples illustrated here, each use the horizon line, a single light source that is above the horizon line and a perpendicular line below the light source with an arrow touching the horizon line. This becomes the vanishing point for the shadows. The dashed guidelines are referencing the light beams from the light source. The solid guidelines from the arrow or viewers vanishing point provide the shadow shape on the ground.

In the first example of the two-point perspective box shown below, the single light source above the horizon line is the point where the light beams start. The dashed lines represent the light beam hitting the corner of the box and going to the ground. Note that the light beams or dashed lines touch three corners of the box. These dashed lines create the guidelines for the shape and size of the cast shadow. The shadow edge lines are created using the two vanishing points used to draw the box.

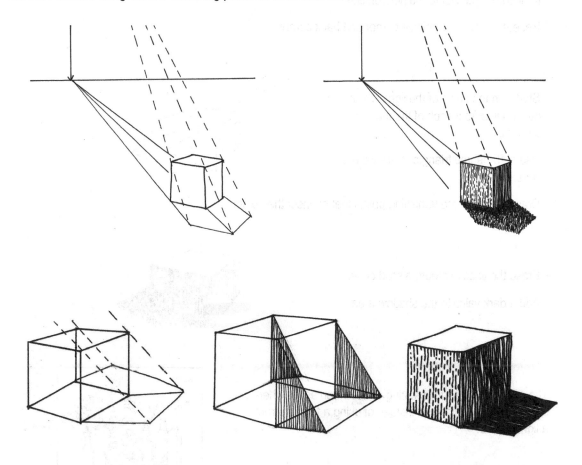

This series of drawings demonstrates how the triangular shapes are used in cast shadows. The first box shows the light beam creating the triangular shape from the box edge to the ground. The second drawing highlights these triangular shapes. The third drawing is a box with gradated value showing shading on the box and shadow on the ground.

PARALLEL LIGHT SOURCE

The parallel method for creating shadows is a shortcut that will imply shadows using parallel lines as the first guideline. This is instead of using a formula from a light source. This method can be used with furniture shadows. This is a simple way to draw shadow is to have it parallel to the horizon line and use a 45 degree angle represent the light ray. Here are the steps you can take to use this technique.

First Drawing

❶ Draw a transparent box in two-point perspective.

❷ Draw the ground shadow lines from the front corner parallel to the horizon line.

❸ Repeat this with the middle corner and back corner.

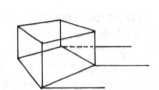

Second Drawing:

❶ Start with the angle of the light, which is a dashed line from each of the three top corners.

❷ Use a 45-degree triangle to draw these lines.

❸ Draw lines from the vanishing points that connect the points.

Third Drawing:

❶ Erase the lines to create a solid cube.

❷ Add a dark value to the shadow area.

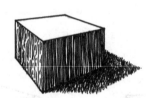

Helpful Hint

The parallel method for creating shadow can assist when showing implied shadows instead of using a formula from a light source.

Chapter 5 – Shade and Shadow

FRONTAL LIGHT SOURCE

As we have been seeing in this chapter, there are factors that will determine the shape of the cast shadow. One is the position of the light source and another is the shape of the object.

In the first illustration, there is the horizon line with the two vanishing points. The four shapes are typical two-point perspective shapes. In the center, the light source above the horizon line becomes the vanishing point for the shadow. The perpendicular line directly below the light source intersects with the horizon line creating the viewer's vanishing point. This point is used to help create shadow on the ground.

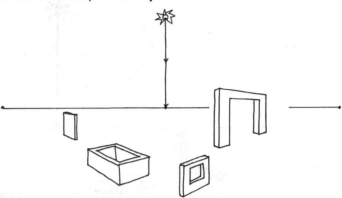

There are three steps in the second illustration for creating shadow shapes on the ground:

❶ Draw dashed guidelines from the light source to the top corner of each object and extend to the ground.

❷ Draw solid guidelines from the point directly below the light source to the ground points of each object.

❸ Draw lines from the left and right vanishing points to form the back line of the ground shadow.

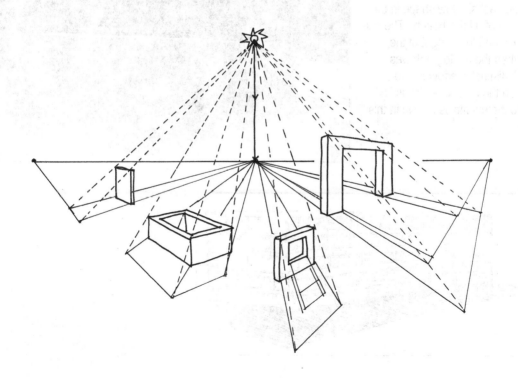

In the third illustration, the guidelines are removed. The shadows are rendered with a dark value. The shading on the objects is rendered a lighter value.

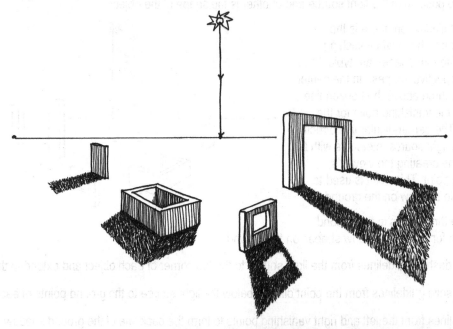

Notice the cast shadow shape in the photograph of a brick bench. The sun was above and to the right of the bench when the photograph was taken. A distinct shadow shape created on the ground, which is similar to the examples shown in this chapter.

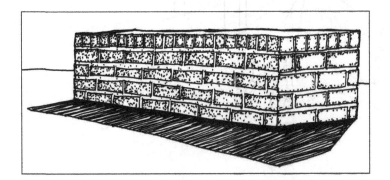

CENTRAL RADIATING LIGHT SOURCE

Similar steps are taken when adding shade and shadow to objects in an interior space. In the illustrations on this page, the light source is a single lamp in the center of a one-point perspective space. The objects are drawn from a single vanishing point on an invisible horizon line

Here are the steps for the second illustration creating the shadow shape on the floor.

① Draw dashed guidelines from the light source to the top corner of each object and extend to the ground.

② Draw solid guidelines from the point directly below the light source, the views vanishing point, to the ground points of each object.

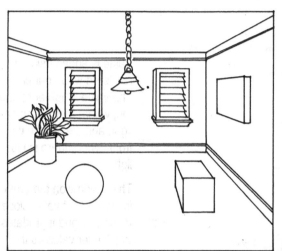
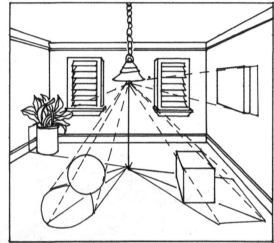

③ Remove guidelines. Render shadows with a dark value object shading with a lighter value.

SHADE AND SHADOW OF PATTERN OBJECTS

When drawing an object that has pattern, you can adjust the pattern value to create shading.

In the box rendering on this page, a single light source establishes the shade and shadow shape. The first drawing illustrates the box shape, the guidelines from the light source, and the resulting shadow shape.

In the third illustration, different values on the each plane depicts shading of the wood rendering. There are also two different rendering patterns used in the image - the wood box is rendered using curvilinear lines and the ground is rendered using parallel lines. Drawing a different pattern for the ground as compared to the object helps the viewer distinguish these two different planes.

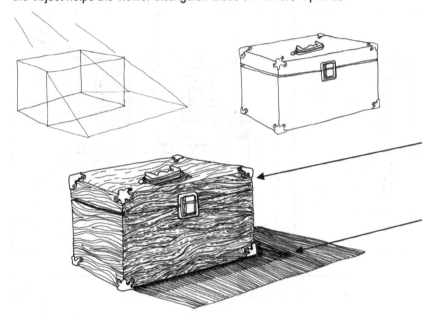

Value is added using a rendered wood pattern. The value is lighter on the top, where there is more light, and darkest on the front, where there is less light.

The shadow on the ground is the darkest value close to the box and it gradates to a lighter value as it moves away from the box.

Helpful Hint

This drawing includes textures and pattern on a wood box, a basket and a ceramic mug. A key aspect of being successful is having different values of dark and light to highlight each object. For example, the basket in the middle of the group needs to stay lighter in value than the mug that is behind it. In addition, the shadow on the left side of the basket is darker than the wood box. When working with a drawing marker you need to start out with light values and add the darker values as you go.

SHAPE AND FORM OF CYLINDRICAL OBJECTS

Graduated values on a cylinder shaped object help to define shape and form. The three examples below demonstrate this technique. The two samples of each shape shows how the first image, with just the contour line, looks flat and without any curved shape. The second examples, with rendered values, have three-dimensional form. The shadow on the table provides a ground for the object and visually reinforces the shape of the object.

Stippling is used to add form and shape with the light source coming from front and center.

In this drawing of the mug, the light source is coming from the left side. To create this illusion shadow was added to the top of the rim and graduated shading was added on the right of the mug and the handle.

This image implies light from the right side by adding shade and shadow to the left. The container candlestick has shadows that graduates in value from light to dark. The container and the book have shading on the ground surface.

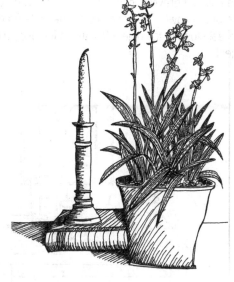

PUTTING IT TOGETHER

When drawing the tabletop of accessories with a light source from a lamp, I worked on three preliminary sketches before starting the final drawing.

Before drawing the objects in one-point perspective, I sketched each accessory using the centerline technique to assist with the proportions.

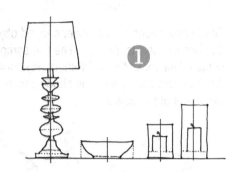

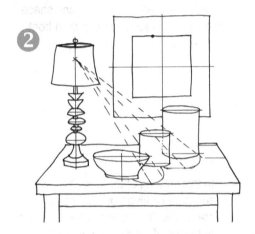

A second quick contour line sketch put the objects together on the table in one-point perspective. This gave me an opportunity to be sure the composition would be successful. At this point changes are easily made if needed. Guidelines from the lamp determine the shade on the objects and the shadow on the tabletop.

The third sketch includes texture and value in the shade and shadow areas of the image. This was another opportunity to practice adding texture and value.

The finished drawing includes the shaded area on the right side of the bowl. The shadows on the table graduate from darkest close to the object to light as it moves away from the object.

FIND YOUR INSPIRATION

As your skills improve, you can begin to expand your drawing subject matter.

In looking for new subjects, think about what you find interesting. Ask yourself, "What objects or topics inspire me?"

Being enthusiastic about the subjects you draw will help keep you focused. Work on finding your own interest and begin to make that the focus of your artwork.

In addition, while you are still at the beginning stages of learning how to draw, it is best to use objects that you can set up in front of you so that you can see proportion, depth, and details. Begin with a single object, and as you gain confidence, add more objects to your compositions.

For example, during my busy everyday life, I enjoy drawing objects that I can gather from around my house. The drawing here is the windowsill above my kitchen where I moved the objects around to assemble a successful composition and then practiced drawing exactly what I saw.

As your drawing skills improve, think about what inspires you.

Part II
Drawing Interior Elements

This next section focuses on drawing interior elements and provides an opportunity for you to apply your drawing skills to communicating design ideas for the interior space.

Chapter 6 is devoted to the drawing of furniture in one-point and two-point perspective. Chapter 7 covers a variety of other interior elements including plants, tabletop objects, accessories and window treatments.

Chapter 6

Furniture

Interior space design by Connie Riik

In this chapter, we will use tables and chairs as models for learning how to draw furniture in one-point and two-point perspective.

Many of the techniques used to draw furniture are similar to those covered in previous chapters. Here you will have the opportunity to expand what you have learned and apply it to drawing more complicated interior design subjects.

GETTING STARTED

Interior designers draw furniture from various viewpoints, such as elevation, top view, one-point perspective, and two-point perspective. Each viewpoint illustrates a different feature of the piece.

An elevation drawing is often the best starting point for drawing furniture. When a furniture piece is drawn in one-point perspective, two-point perspective or top view, there is no measurable scale to help guide the proportions of the drawing. However, when a furniture piece is drawn in elevation view, the drawing includes a measurable scale of the object and its unique features. Therefore, if you begin with an elevation drawing, you can then use it as a tool for drawing furniture with correct proportions in the various perspectives.

Scale is a drawing method used to represent the true dimensions of an object. For example, in elevation drawings, it is often most effective to use the scale: ½ inch = 1 foot.

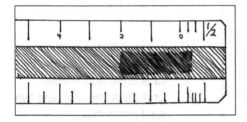

Using a ruler, you can draw an object to scale, making it smaller or larger than the real object.

BEGINNING WITH THE BOX

For many pieces of furniture, you can begin with a basic box. For example, the bench below is an altered version of a box.

To create the bench, tape tracing paper over a box drawing and trace the lines of the box. Then, adjust the box on the tracing paper to create a bench. Later in this chapter, we will work through these steps in more detail.

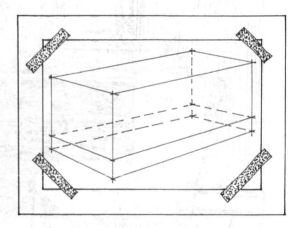

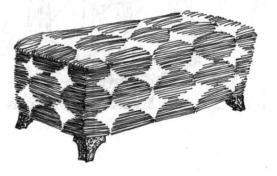

FURNITURE ANATOMY

To differentiate pieces of furniture in your drawings, notice each one's distinctive features.

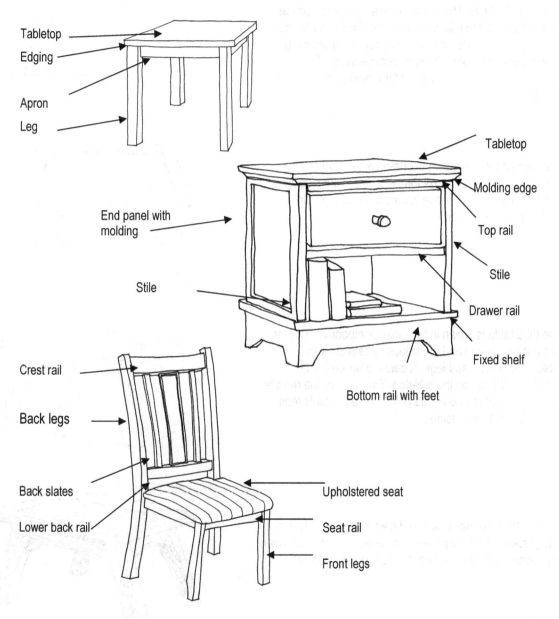

Tabletop
Edging
Apron
Leg

Tabletop
Molding edge
Top rail
Stile
Drawer rail
Fixed shelf

End panel with molding

Stile

Bottom rail with feet

Crest rail
Back legs
Back slates
Lower back rail

Upholstered seat
Seat rail
Front legs

Refer to this page as you progress through the chapter.

DRAWING DIFFERENT VIEWPOINTS

In previous chapters we focused on drawing the box from various views. Now we can translate what we have learned about the box and apply it to drawing furniture.

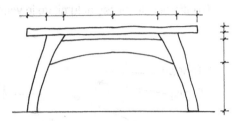

In the row of tables on the right, the first one is an elevation drawing of a table. The notched lines help guide you as you draw the table's features to scale. From an elevation drawing like this one, you can continue on to draw the same piece of furniture in other perspectives. The elevation drawing will provide information about scale and proportion.

The second table is drawn in one-point perspective. The front of the table is a flat rectangle. The perspective lines that form the tabletop move toward a vanishing point centered behind the table.

The third table is drawn in two-point perspective. The table has a leading edge and guidelines to help draw the tabletop and recessed legs. A detail to notice in this drawing is the rim on the tabletop. The edge of the rim sits on the same guideline as the point where the table legs recess in from the outside.

Finally, this table has been rendered with a wood texture and pattern. The front plane of the table has a darker value indicating a light source from the top back left corner.

CREATING AN ELEVATION DRAWING

The key to drawing realistic furniture is correct proportions and details. One method to accomplish this is to begin with a scaled elevation drawing.

For example, follow these steps to see how to draw an elevation of the table below. The scale for this drawing is ½ inch = 1 foot. *(The pictures below are not to scale.)*

❶ Begin with a light pencil drawing of the center box shape.

❷ Next, draw the fixed shelf below the box. Notice the guidelines on each side show the wider width of the shelf.

❸ Then, draw the bottom rail, including the legs. Notice this section has the same width as the original box.

❹ Now, use the side guidelines to draw the tabletop and molding edge. This should be the same width as the lower fixed shelf.

❺ Add the top rail and drawer rail.

❻ Add the center of the box with the two side panels, the top panel, and the drawer.

❼ Locate the center of the drawer by drawing an X from corner to corner. Draw a circle for the knob at the center point of the X.

❽ Finally, trace the pencil drawing with markers. Use a wide marker for the outlines and a finer marker for the interior details.

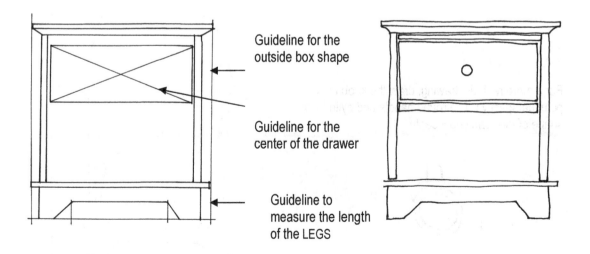

Guideline for the outside box shape

Guideline for the center of the drawer

Guideline to measure the length of the LEGS

DRAWING ONE-POINT PERSPECTIVE FURNITURE

To draw furniture in one-point perspective, begin with a front elevation drawing. Using measurements obtained from the elevation drawing, add the perspective's depth. Follow these steps to see how to draw the table below in one-point perspective.

❶ First, establish a horizon line above the top line of the drawing. The horizon line should be the same length as the height of the table.

❷ Next, locate the center of the width of the table. Make a dot on the horizon line here. This is the vanishing point.

❸ From each corner of the tabletop, draw a perspective line to the vanishing point.

❹ To draw the back of the table, draw a line that is half the height of the table and parallel to the front edge.

❺ To draw the drawer knob in one-point perspective, draw the stem of the knob at its center and the round pull slightly below.

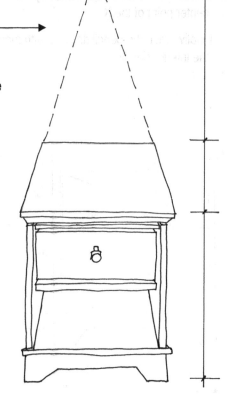

The tabletop perspective lines lead back to the vanishing point.

To draw the inside shelf, draw two perspective lines from the bottoms corners of the shelf to the vanishing point.

Helpful Hint:

For a more realistic drawing, draw the knob in one-point perspective. The round handle and cylindrical shape of the stem show depth.

DRAWING UNIQUE FEATURES

Many furniture pieces have unique features and details that may take several steps to add to your drawings. Begin with a scaled elevation drawing to show the shapes and proportions of the furniture piece. Then continue by using the outside box shape as a guide for locating various key parts of the piece.

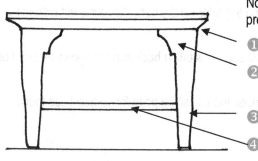

Note the unique features of the table to the left in this preliminary scaled elevation drawing.

1 The top with a molded edge extends out from the base.

2 Instead of an apron, the added molding of the legs provides support for the tabletop.

3 The legs are tapered, narrowing toward the bottom.

4 There is a lower shelf.

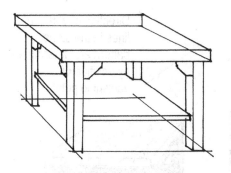

Notice in this first one-point perspective sketch, the box shape is drawn in light pencil and most of the unique features are blocked out. These include a tabletop larger than the base, the shape of decorative molding on the legs, and the start of a lower shelf. Note that the lower shelf hides the bottom of the back corner leg, and thus that portion of the leg was left out of the drawing.

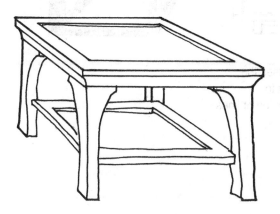

The final drawing, done in marker, shows all the table's unique features in more detail, including the leg molding, the bottom shelf and the tapered legs.

CASE GOODS IN ONE-POINT PERSPECTIVE

When drawing case goods in perspective it is important to draw the legs correctly by using perspective guidelines. The location of the furniture piece to the vanishing point will determine what side of the legs is visible. In the illustration below, each case good drawing is at a different location relative to the vanishing point. In the drawings below, notice the following:

1 The table on the left shows the right side of the legs and the table on the right shows the left side of the legs.

2 The middle table shows a flat front drawer with the shelf above receding back with the vanishing point as a guide.

3 The wood and stone patterns on the top of the tables uses the vanishing point for vertical lines found in the pattern.

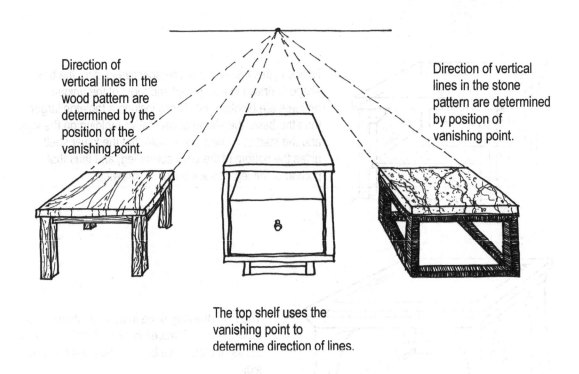

Direction of vertical lines in the wood pattern are determined by the position of the vanishing point.

Direction of vertical lines in the stone pattern are determined by position of vanishing point.

The top shelf uses the vanishing point to determine direction of lines.

ADDING SHADOW WITH PARALLEL LINE METHOD

A simplified method of adding shade and shadow to a one-point perspective furniture piece is to use the parallel line method. The shadow is created with parallel lines and 45-degree angles using steps outlined below:

1. Draw the ground shadow lines from the front corner leg parallel to the horizon line.

2. Repeat this for the middle corner and back leg.

3. Add the angle of light depicted by a dashed line from each of the top corners. Use a 45-degree triangle to draw these lines.

4. Draw lines connecting the intersections of the ground lines and light angle lines.

5. Add shade and shadow using varying values.

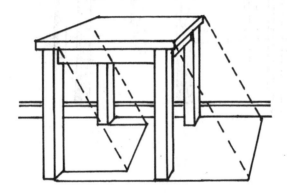

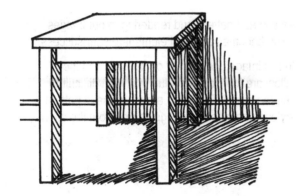

 Table Shade

Floor Shadow

Wall Shadow

ADDING SURFACE RENDERING

Rendering is the act of adding value, shadow and interest to drawings. It is important to remember as you render drawings that furniture is viewed from a distance. This will affect how you see pattern and thus how you should draw it. The concept of foreshortening, as discussed earlier in the book, also applies to patterns on objects.

To render foreshortened pattern, you can create this illusion by drawing the pattern with less detail as the piece of furniture recedes from the viewer. The drawing below is an example of adding wood texture to a side table.

Compare the small wood pattern sample to the wood rendering on the table. The rendering on the table is less detailed, simplified and has smaller value changes.

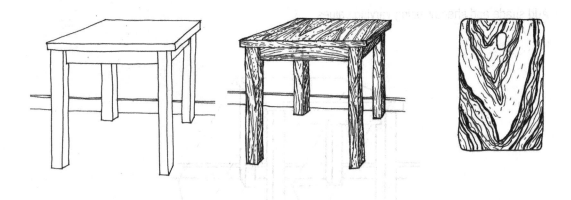

Helpful Hint:

As you add pattern and rendering to our images value becomes a critical element to your success.

In this image, the pattern on the sofa and the pillow are too close in value making it difficult to see the pillow. They also are too close in pattern styles to be able to distinguish one from the other.

SIDE CHAIRS IN ONE-POINT PERSPECTIVE

To draw a chair in one-point perspective, center the vanishing point behind the chair.

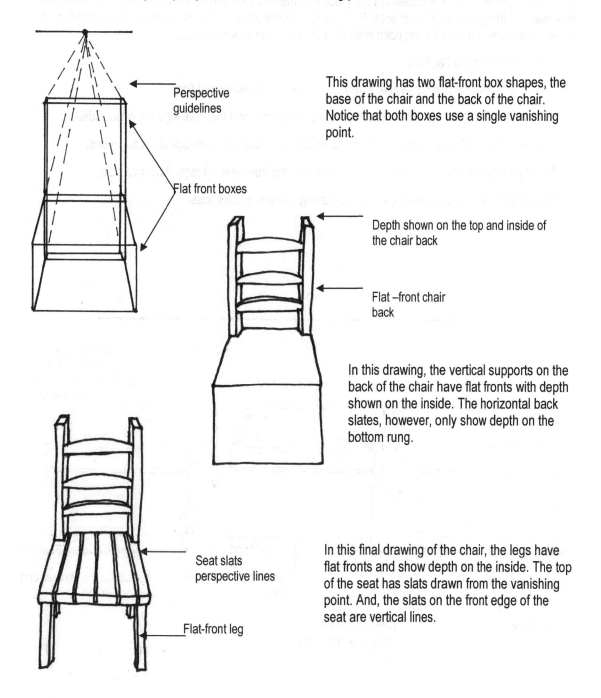

Perspective guidelines

Flat front boxes

This drawing has two flat-front box shapes, the base of the chair and the back of the chair. Notice that both boxes use a single vanishing point.

Depth shown on the top and inside of the chair back

Flat –front chair back

In this drawing, the vertical supports on the back of the chair have flat fronts with depth shown on the inside. The horizontal back slates, however, only show depth on the bottom rung.

Seat slats perspective lines

Flat-front leg

In this final drawing of the chair, the legs have flat fronts and show depth on the inside. The top of the seat has slats drawn from the vanishing point. And, the slats on the front edge of the seat are vertical lines.

OTTOMAN IN ONE-POINT PERSPECTIVE

In one-point perspective, the location of the piece of furniture in relation to the vanishing point determines how much of that piece the viewer sees. For example, the ottoman below, drawn three times in different locations relative to the vanishing point illustrates the following characteristics:

1 Each ottoman has a flat front.

2 Each line is either a horizontal line, a vertical line, or a vanishing point line.

3 The center ottoman does not show either side of the piece, and the front legs show two sides.

4 The left ottoman shows the front and right side, and the front and right legs show two sides.

5 The right ottoman shows the front and left side, and the front and left legs show two sides.

6 The buttons on the top of each ottoman rest along vanishing point lines.

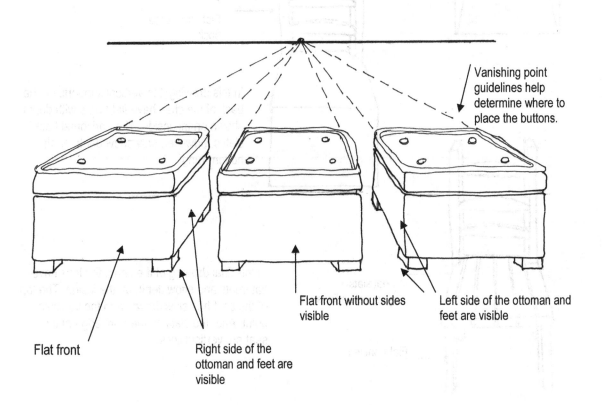

Vanishing point guidelines help determine where to place the buttons.

Flat front without sides visible

Left side of the ottoman and feet are visible

Flat front

Right side of the ottoman and feet are visible

PATTERN IN ONE-POINT PERSPECTIVE

In one-point perspective, the location of the piece of furniture in relation to the vanishing point determines how the viewer sees the pattern and the furniture shape. The example below shows the ottoman patterns and furniture shapes in perspective.

1️⃣ The pattern of the material on the flat front is to scale and there is not any perspective distortion.

2️⃣ On the top of the ottoman, the stripe and zig-zag patterns are drawn using the single vanishing point as a guide.

3️⃣ The pattern on the sides of the ottoman uses the vanishing point and vertical lines. The side pattern will get shorter because they are foreshortened.

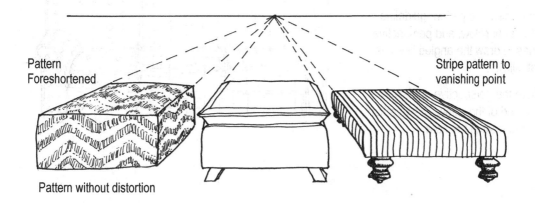

Pattern Foreshortened

Pattern without distortion

Stripe pattern to vanishing point

Here is an example of viewing chairs from different points of view from the single vanishing point.

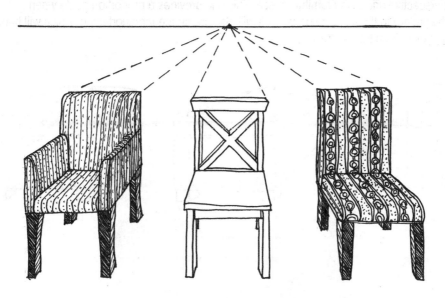

FABRIC PATTERN ON A SOFA

Here is a one-point perspective sofa that demonstrates steps for adding a fabric that has horizontal and vertical pattern. It is helpful to use a grid on the sofa to provide guidelines for creating the sofa pattern.

To add the guideline grid:

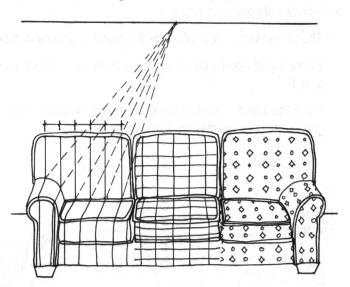

❶ Start on the left top corner of the sofa back and draw vertical lines that are equidistance apart.

❷ Draw vertical lines to the bottom where the back pillow meets the seat pillow.

❸ From where the pattern guideline meets the seat pillow, add perspective guidelines to draw the angled lines on the seat pillow.

❹ Where the perspective guidelines meet the end of the pillow, draw vertical lines on the front of the sofa.

Helpful Hint

To add a fabric pattern to a furniture drawing, start by drawing the pattern's design on a grid as shown in the inset. This will give you the proportions for pattern design.

Next, add a perspective grid to the furniture piece. This grid provides a proportion guide when transferring the pattern to the furniture drawing. It will also create the foreshortening, which will happen as the pattern moves from front to back.

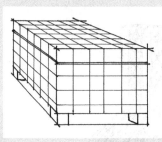

VARIETY OF PATTERN AND VALUE

One of the key aspects of adding pattern to your rendering is to have a variety of pattern and value. This is true when choosing fabrics for an interior space. Contrasting the pattern and value adds interest to the interior space.

In the interior line rendering below, contrast exists between the sofa pattern and the different pillow patterns. The sofa pattern, which is a light valued, small geometric pattern, is drawn first. Next, each pillow has a different type of pattern and a darker value than the sofa. The different patterns include a geometric strip, a crisscross horizontal stripe, a diagonal and floral pattern as depicted on the bottom row of the fabric examples.

ONE-POINT FURNITURE STEP-BY-STEP REVIEW

This series of drawings reviews the process of drawing an arrangement of furniture with accessories. It is in one-point perspective with the single vanishing point.

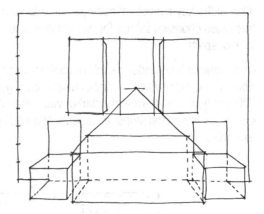

The first drawing starts with the large rectangular shape representing the ceiling and wall lines. In a one-point perspective, the back wall is an elevation that has a scale. There are dimension lines noted on the left side of the wall. This provides guides for the height of each object. The same dimensions are used on the bottom floor line to provide the width guides for each object. The furniture and accessories start with a flat front, using the vanishing point for defining perspective lines.

In the second drawing, the shapes have been refined and details added. A plant and container were added on the cabinet top, drawers and doors were added to the cabinet, the chairs were defined and depth was added to the paintings.

In the third drawing, pattern was rendered on the floor, cabinet and chairs. A variety of techniques using line and stippling was used to create the rendering.

SKETCH PRACTICE PAGE

Sketching one-point perspective tables and adding objects to the drawing can be a simple way to practice your drawing skills. I will either make up a combination of table and objects or be looking at an existing table and use this as a guide for assisting with the proportion of each piece.

In each of these drawings, a vanishing point was placed above the middle of the tabletop to create guidelines for sides of the table. Using the table size as a guide for proportion, the table legs and drawers were added. The centerline technique is used add a lamp. Other accessories are included such as plants, picture frames or art on the wall.

The two drawings below include shade and shadow. In each drawing, the light source is coming from the right side. The objects on the table have a darker pattern on the left and a dark area on the left for the shadow. The shadow below the table was created using parallel shadow method.

DRAWING TWO-POINT PERSPECTIVE FURNITURE

To draw furniture in two-point perspective, begin with a scaled elevation drawing. From the elevation drawing, then add the perspective's depth. The following steps explain how to draw the table below in two-point perspective:

❶ Start with a leading edge and the center box shape.

❷ Next, add the tabletop with the molding edge extending out from the center box shape and the bottom shelf, base rail and feet.

❸ Add drawer perimeter lines and draw an X from corner to corner to find the center. Place the knob at the center of the X.

❹ The image is finished by adding end panel molding and the inside depth line of the shelf.

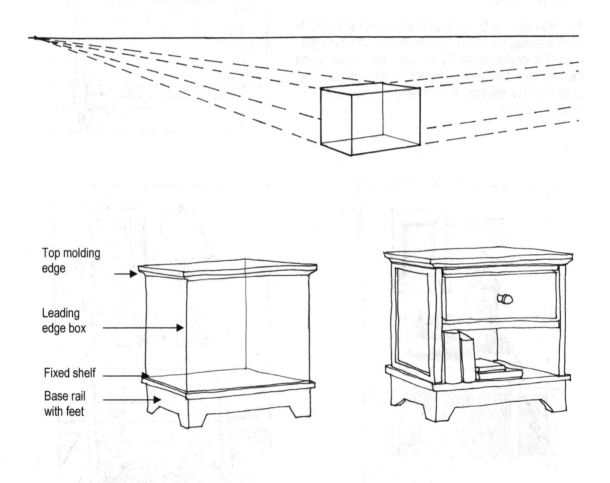

Top molding edge

Leading edge box

Fixed shelf

Base rail with feet

PLANES OF TWO-POINT PERSPECTIVE FURNITURE

One of the things to remember when drawing two-point perspective furniture is that there are two parallel planes on each piece - a left parallel plane and a right parallel plane. The bookcase and table illustrate this concept with a light value plane, a dark value plane and a white top or bottom area. The left side of the table uses the left vanishing point to determine the angle of legs and tabletop. The lighter diagonal lines depict this left parallel plane. The same concept is applied to the right parallel plane, which is illustrated by the darker value.

Both the table and the bookcase have step-by-step drawings illustrations later in the chapter.

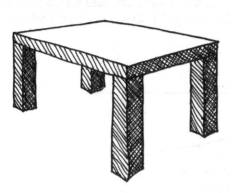

The lighter value represents the left parallel plane. The left vanishing point is used.

The darker value represents the right parallel plane. The right vanishing point is used.

This concept is especially helpful to keep in mind when drawing a two-point perspective bookcase. This illustration shows the lighter value representing the left parallel planes and the darker value representing the right parallel planes.

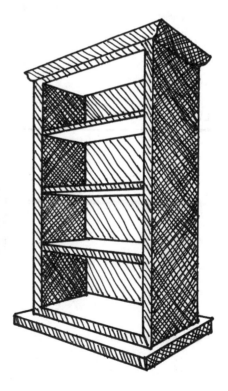

TABLES IN TWO-POINT PERSPECTIVE

These two illustrations demonstrate the steps for drawing the legs on the table in two-point perspective. Note the second vanishing point is off the page and the dotted lines are perspective guidelines from this perspective point.

The first drawing illustrates how the legs are positioned using the two vanishing points. This was drawn first and shows the perspective guidelines creating the table leg shape on the ground. Each leg has two perspective guidelines from each vanishing point. It can be helpful when you are first drawing two-point perspective legs to take the time to draw these guidelines. In the future, you will be able to add these without guides

The second illustration shows a finished table. Notice how the darker side plane uses the right vanishing point (RVP) and the lighter side plane uses the left vanishing point (LVP). This is true for the sides of the legs that are underneath the tabletop.

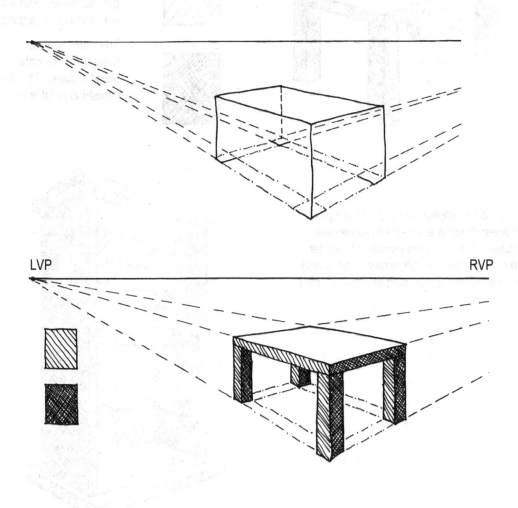

LVP   RVP

TABLE SHADOW IN PERSPECTIVE

Here is a detailed process for of using the parallel method to draw shadows for a two-point perspective table.

First Drawing:

❶ Draw the ground shadow lines from the front corner leg parallel to the horizon line.

❷ Repeat this with the middle corner leg and back corner leg. Notice the bottom of the leg is drawn to provide this corner.

❸ Start with the angle of the light, which is a dashed line from each of the three top corners.

❹ Use a 45-degree triangle to draw these lines.

❺ Draw lines connecting the intersections of the ground lines and light angles.

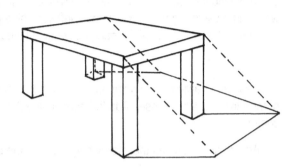

Second Drawing:

❶ Draw the ground shadow lines from the bottom corners of the three legs.

❷ Draw a center guideline from the left side leg.

❸ Connect the intersecting points from the ground shadow lines to the edges of the top shadow.

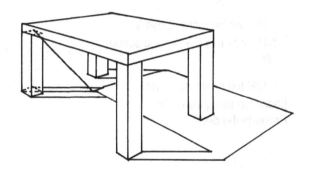

Third Drawing:

❶ Erase the extra guidelines.

❷ Add a dark value to the ground shadow area.

❸ Add a lighter value to the shaded area on the table

Shade

Shadow

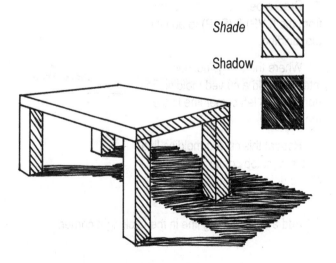

BOOKCASE IN TWO-POINT PERSPECTIVE

Here are steps demonstrating drawing of a two-point perspective bookcase. The finished image to the left includes accessories and a plant.

The drarwing was started by setting up a typical horizon line (HL) with a right vanishing point (RVP) and a left vanishing point (LVP). The dashed lines represent the perspective guidelines.

First Drawing:

❶ Begin by drawing a two-point perspective box that will be the outer edge of the bookcase. Add the horizon line (HL) through the center.

❷ Add a second line as the interior edge around the box.

❸ Using the "X" method, find the middle and add two lines representing the middle shelf using the left (LVP).

❹ Repeat the steps above to locate and draw the top and bottom shelf.

❺ Add the right side of the bookcase using the (RVP). Estimate the depth.

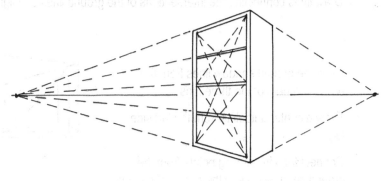

Second Drawing:

❶ Draw a perspective guideline from the (LVP) & (RVP) to add the top molding.

❷ Where the two guidelines intersect, add a curved molding line from the top left corner line to the molding line.

❸ Repeat this curved molding line on the right top corner. Use your eye to determine the length of the molding

❹ Add a third molding line in the back right corner.

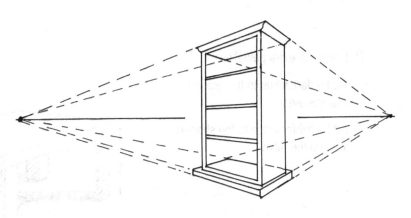

⑤ To add bottom molding, draw two parallel lines from the RVP and two parallel lines from the LVP. Where they intersect, draw a horizontal line representing the bottom right front corner.

⑥ To add the bottom base, first draw the inside line using a perspective line from the RVP to the left front corner of the bookcase. Draw a second perspective line from the LVP to the right corner.

Third Drawing:

❶ Draw a vertical line from the inside top left corner down to the bottom inside left corner.

❷ Use the perspective guidelines from the shelf front to the RVP to determine the shelves.

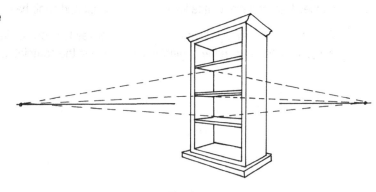

In Summary:

This drawing illustrates the two parallel planes as noted in the beginning of the chapter.

The left vanishing point is used to draw the vertical lines for the plane with the lighter value.

The right vanishing point is used to draw the vertical lines for the plane with the darker value.

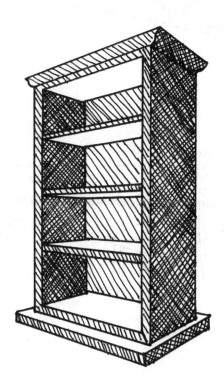

SIDE CHAIRS IN TWO-POINT PERSPECTIVE

One way to view a chair in two-point perspective is as two boxes fit together. Therefore, to draw a two-point chair, proceed as follows:

① Start by drawing a leading edge box that forms the base of the chair.

② Then, add perspective lines and vertical lines to complete the two box shapes that form the foundation of the chair back and chair base.

③ As the second drawing illustrates, refine the boxes to show the legs, seat and chair back. Notice the original leading edge in the completed drawing.

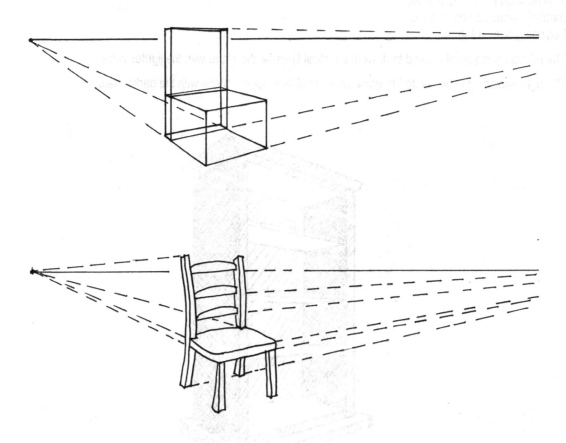

OTTOMAN IN TWO-POINT PERSPECTIVE

The example below shows an ottoman drawn in two-point perspective. Note the following characteristics:

❶ The ottoman has a leading edge.

❷ Each line is either a vertical line or a vanishing point line.

❸ Each leg of the ottoman shows a left and right side.

❹ The buttons on the top of the ottoman rest along perspective guidelines are drawn from the vanishing points. The legs show the left and right sides of the feet.

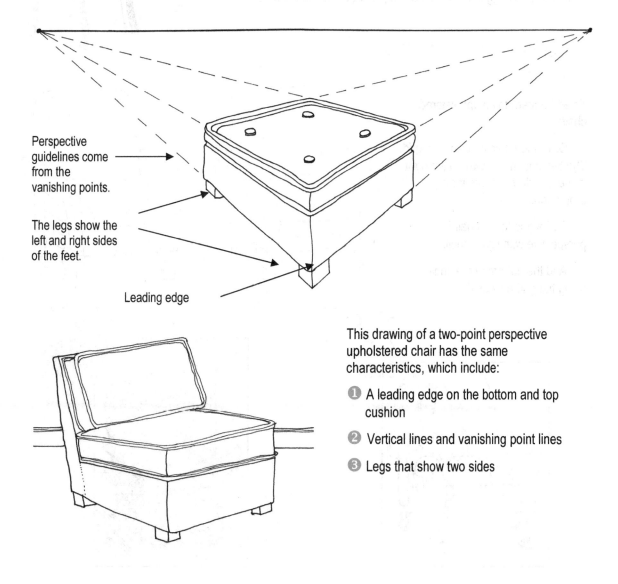

Perspective guidelines come from the vanishing points.

The legs show the left and right sides of the feet.

Leading edge

This drawing of a two-point perspective upholstered chair has the same characteristics, which include:

❶ A leading edge on the bottom and top cushion

❷ Vertical lines and vanishing point lines

❸ Legs that show two sides

ADDING SURFACE RENDERING

The model used for this drawing is a wood chair that is painted black with an upholstered seat. Here is an example of how a drawing of this chair could be rendered:

❶ For the wood finish, the darkest value of each left side panel was drawn with vertical, horizontal and crossed lines. The medium value on the right side panels uses stippling.

❷ The upholstered seat fabric is actually an intricate striped pattern, but is rendered with suggested stripes of simple lines and small swirls.

ADDING FABRIC PATTERNS

To add pattern to an upholstered chair:

❶ Draw the fabric pattern on a grid. The two patterns drawn on the right in the grid demonstrate their proportions.

❷ Add a grid to the chair in perspective with light pencil.

❸ Add the pattern to the chair using the grid as a guide.

TWO-POINT FURNITURE STEP-BY-STEP REVIEW

Now that we have gone over drawing furniture in detail, here is a quick step-by-step recap of the basic steps for drawing this upholstered bench.

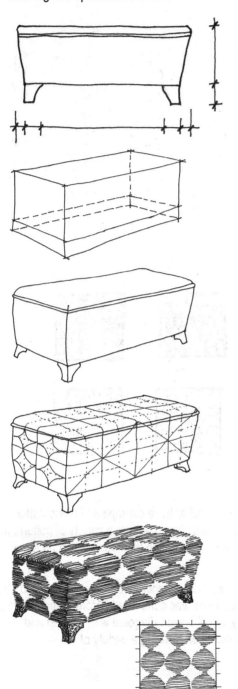

❶ First, draw a scaled elevation of the bench. Note how the sides of the bench slope inward and how the feet point outward.

❷ Draw a two-point perspective box using the proportion from the elevation drawing.

❸ Next, layer a sheet of tracing paper on top of the box drawing and refine the curved shape of the top of the bench, the sloped sides, and the outward-pointing feet.

❹ Now, layer a second sheet of tracing paper on top of drawing number two. This will help establish the gridlines to add a fabric pattern to the bench. To make the grid, start by drawing an X on each panel of the bench from corner to corner to find the centers.

❺ Place a third sheet of tracing paper over the grid, and draw the fabric pattern in pencil.

❻ Finally, finish the drawing by going over the pencil lines in ink and adding any remaining details.

SKETCH PRACTICE PAGE

As you are adding fabric patterns to your furniture pieces, it is helpful to have examples of fabric patterns to use as a reference. Take time to collect a group of actual fabric samples that have a variety of pattern. In your sketchbook draw a sample of each pattern. This will give you the option to choose which pattern will work best in your image.

It was fun to play around with a variety of patterns in this sketch. I cut a bunch of different patterns so I had actual samples to draw from. A quick sketch of five distinct patterns was completed before putting them on the pillows in the final sketch. The patterns include a paisley, small flower, diagonal line, zig-zag and checker. As I added the patterns to the sofa pillows, I also worked on drawing a variety of value.

PUTTING IT TOGETHER

This image is from a design created by Connie Riik and provides a good representation of many of the elements covered in this chapter.

The design has a wide range of value from dark to light. There is a dynamic contrast of value between the light fabric of the chair and the dark value of the wood. There is also a variety of finishes. Typically, interior spaces are designed to create interest by including contrasting values and a range of finishes with various textures and patterns.

I enjoyed rendering this image and used a variety of lines and stippling pattern to represent the finishes.

Design by Connie Riik

A. Cabinet wood	B. Rug pattern	C. Wall color	D. Chair fabric	E. Flooring wood	F. Lamp

A. Cabinet is darkest value, very little white is showing with curved lines to imitate wood grain.

B. Rug has curves in the pattern. I imitated the shapes that were found in the boarder and main area of the rug. Stippling was added to provide a general value for the rug.

C. Walls were painted a medium to light blue grey and I used a vertical close line pattern.

D. Chair fabric was a solid white woven fabric. I used stippling to represent the shading on the chair.

E. Flooring wood pattern is similar to the cabinet wood and yet it is lighter in value.

F. Lamp shade was white and I left this without any rendering.

Creativity Strategy
CREATIVE CHANGES

One of the common obstacles while learning to draw is the tendency to stop simply because you made a mistake. I would like to suggest that you reframe your perceived mistake as a creative change.

What is a mistake anyway? For me it is when your drawing looks differently than what you had planned or expected. You expected your drawing to look more like the object you are drawing, and when it does not, you call this a mistake. However, a fundamental part of the drawing process is being able to make adjustments and changes as the drawing develops. If something you draw does not fit in with your original plan, instead of stopping, try to revise, regroup and readjust your drawing. This is a creative change!

I have found that I get better at being flexible and making adjustments the more I practice drawing. One method you can try is to work on drawing for about half an hour and then stop and walk away. When you return, you will be able to look at the drawing with fresh eyes.

Now you can regroup and readjust to make creative changes. I find that my final piece is more successful when I allow myself to shift and change as I am working. Also, remember, your drawing is an artistic expression. It is not a photograph. Often these creative changes can add charm and character to your image.

Try it. Allow yourself to stop and see what you have drawn. Ask yourself if there are any adjustments, revisions, or creative changes that you can make.

Chapter 7

Interior Details

Interior space design by Connie Riik

Interior spaces are designed for people to enjoy and to feel comfortable being in them. Your interior drawings will be more interesting when you add personal flavor such as accessories, plants, artwork and window treatments. Frequently, these elements are referred to as entourage.

In this chapter, we will look at techniques for successfully incorporating the humanizing element into your drawings.

GETTING STARTED

Adding interior details to your drawings is an exciting way to make them interesting. Accessories, everyday objects and figures can all be added to your interior drawings.

A successful composition of these items is an important consideration. Composition in design refers to the placement or arrangement of visual elements or ingredients to create a sense of unity. The goal is to have each of these separate items put together as one unified form. Here are a number of key principles of composition to keep in mind to create an effective design of interior elements in your drawings. This image highlights each of these principles.

Composition techniques. There are several guidelines for creating pleasing combinations of visual elements. One important technique is the "rule of odds" which suggests that an odd number of elements in an image is more interesting than an even number.

Balance. Refers to the ways in which the elements are arranged. On the desk top, the tall lamp on the left is balanced with the basket on the right. In addition, each side of the desk has an object on the floor. This is creating an approximate symmetrical balance.

Contrast. The difference between the darkest and lightest values refers to contrast. Here, contrast of values is shown by the dark value of the wainscot against the light value of the desk and plant. This visually distinguishes the different objects and provides interest.

Variety. Different shapes, textures and size add interest to the composition. Here, the variety is shown with the two textures of basket weave, fabric pattern on the chair, wood texture on the floor.

Overlap. Refers to the layering of objects to show depth and distance. Here, the framed picture was drawn to be behind the lamp, the chair in front of the desk and the leaf in front of the desk all provide visual cues for showing depth.

Visual Weight. When comparing items, those that are solid and larger than a smaller and darker item, may appear similar in same size.

BALANCE

Although all the principles are significant in creating an effective design, the composition needs to appear balanced. As you arrange objects, imagine a line through the center of your page. The goal is for each side to appear equally balanced by distributing negative and positive elements. You can achieve balance by arranging shapes in such a manner that their **visual weight** combines to create a sense of unity.

Negative shapes represent areas of emptiness. Positive shapes take up space. In the two images below, the negative space is the area around the table and accessories. The original drawing had more negative space that was cropped out to create a more spatially balanced image.

A simple way to look at this is be sure one side does not look heavier that the other. Keep in mind, balance can be either symmetrical or asymmetrical.

SYMMETRICAL BALANCE

With symmetrical balance, visual balance is achieved when each side of a drawing mirrors the other. Notice in this image, if it is cut in half, objects match on either side. The clock is symmetrical and there are matching plants and candlesticks on either side. This type of **formal balance** provides an easy solution for creating balance.

In the interior design field, approximate symmetrical balance is also used. This is when your first impression is that of symmetry. Weight may be identical but not a mirror image.

ASYMMETRICAL BALANCE

With asymmetrical balance, each side of a drawing contains different elements, but the sides have equal visual weight. This **informal balance** creates a visually more dynamic image.

The adjacent drawing demonstrates these techniques for creating an asymmetrical balance:

1. Use multiple objects on one side, which provide more weight, and a single object on the other.

2. Use darker values to achieve more visual weight.

3. Use objects with contrasting shapes to establish visual weight. For example, notice the plant as compared to the pictures and clock.

ADDING ACCESSORIES

TABLE TOP

Generally, there are only several items on a tabletop. This often includes a lamp. The height of the lamp is an important consideration with the overall composition. The accessories add a variety of shape and texture. This image uses asymmetrical balance to create a sense of cohesiveness. The tall lamp on the left works effectively with the size and shape of the picture. The objects on the table provide visual weight.

BOOKCASE

Bookcases will hold many different types of usable and decorative items. When creating balance and interest in a bookcase these are factors to consider:

❶ Each self needs balance and a variety of different textures and shapes.

❷ The overall bookcase needs balance.

❸ Visually heavier items often work best at the bottom of the bookcase.

❹ Consider using framed pictures and plants. Stack books horizontally or vertically.

Helpful Hint

Before starting to draw accessories in your bookshelf, take time to play around in your sketchbook. Draw a group of shelves and fill each shelf with different combinations of accessories. Use this sketch as a visual reference for your final drawing for ideas.

ARTWORK

Artwork is another important feature to add to your perspective project. As you begin to incorporate artwork into your piece, remember to:

1. Create a vague, low-value image to keep the artwork from becoming an unwanted focal point.

2. Find an image that is inspiring to you. Lightly sketch an impression or suggestion of the image.

3. Avoid using symbols that attract the eye and create an over emphasized focal point, such as triangles, crosses, or spirals.

4. Avoid drawing large, single objects, such as a flower, a tree, or a shell.

5. Keep the image shapes and lines in a low value to avoid attracting the viewer's eye with contrasting values.

This example of a subtle landscape of mountains and sky was rendered with stippling to give the impression of the shapes and show small changes in value. Notice there is little contrast in the painting. This prevents the artwork from becoming a distraction to the overall interior drawing. The goal for any piece of artwork included in an interior drawing is for the image to be a part of the background.

LAMPS

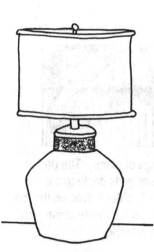

Lamp bases and shades come in a large variety of shapes and sizes. To draw the correct proportion of the lamp base and shade it is important to have their actual dimensions. I am often surprised at how wide and tall a lampshade actually is.

This illustration shows the different dimensions that are used to draw the proportions of the lamp and shade.

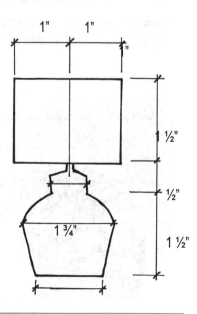

DRAWING PLANTS

Plants add an organic subject to your drawings. Although plants have a free-flow design, there are techniques for drawing them realistically. The first step is to practice looking at a plant as you draw, noticing its unique characteristics. In this section, we will learn how to sketch leaves, stems and flowers and then put each element together using skills covered in previous chapters.

Drawing complicated objects can appear overwhelming at first. Breaking objects down into smaller components is the key moving forward. Throughout this section, we will learn how to draw plants using this method. You can then utilize this method to draw other complicated subjects. Taking time to see a subject in smaller parts as you begin sketching can lead to a more successful finished piece.

To get started, we will focus on the following individual parts of a plant and their unique qualities:

Leaf. Extension of the stem of a plant, its function being to assist with providing food material. You will find that leaves have unique shapes and qualities that will be a part of your drawing.

Vein. Vessels which transport food from the root through the stem into the leaf. Drawing leaves includes the veins.

Bud. The beginning of a leaf or flower found on the end of a stem. In your drawing, adding buds provides contrast of size.

Flower. An esthetic feature on a plant that provides the reproductive apparatus. Drawing plants with flowers provides shape, texture and value variety in addition to the leaves and stems.

Stem. Ascending axis of a plant from which leaves and flowers develop. Stems have unique shapes and qualities that will be included in the details of your drawings.

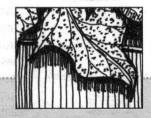

Helpful Hint

Using value will enhance your drawings of plants. The darker value under the leaves on the container adds depth to the drawing. Having a contrasting value of lighter leaves on the top outside of the plant and darker leaves on the inside of the plant will also add depth and interest to the drawing.

THE LEAF

Each plant has a distinct leaf shape and characteristic. Drawing realistic plants successfully starts with looking closely at the unique characteristics and qualities of the leaves so they can be incorporated into the drawing.

Leaf shape:

Notice the distinct outline that defines the shape of the leaf.

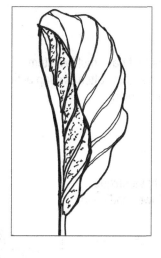

Leaf Value:

One way to add value to a leaf is to distinguish the different sides of it by creating darker and lighter areas. On this leaf, darker stippling gives the back and the bottom sides of the leaves.

Leaf Pattern and Texture:

Look for patterns on each leaf. On the leaves above, notice the outside boarder, the thick vines and curved lines.

Texture helps give leaves and the plant their overall definition and character.

LEAVES

Now, focus on drawing multiple leaves together. As you are drawing, look for overlapping leaves, size varieties and their connection to the stem.

Leaves overlap. In order to draw this realistically, begin sketching in pencil. With pencil, you can draw the overlap, and then use your drawing marker to ink over the pencil lines you want to keep and erase the rest of the lines. Draw the leaves in front first and then add the leaves in back.

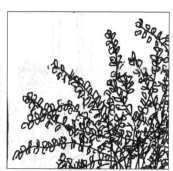

In this fern, there were very small round leaves from the top to the bottom of the stem. Notice how they connect.

STEMS

Like leaves, stems have their own unique characteristics.

Multiple Stem Base

In this plant, multiple stems clump together, originate at the base, and have leaves that grow off each stem.

Stem Texture

Stem texture is unique to different types of plants. In your drawings, imitate the type of stem texture that you see.

Stem Overlapping

In plants that have multiple stem bases, the stems often overlap at the base of the plant.

FLOWERS

Just like leaves and stems, flowers also have unique characteristics. Look carefully at each flower and try to imitate the details in your drawing.

The bud, combined with the other views of the flower, creates interest and contrast of size.

PLANTS STEP-BY-STEP

Sketch the basic contour of the plant in pencil to show different aspects of the leaves and stems. Begin with the stem, then the front leaves, then the back leaves that are overlapping. Finally, add the outline of the container. The pencil drawing should provide a base as you begin to add to the piece. Your finished drawing will be more successful if you begin with an outline before you add the details.

Once you have an outline in pencil, begin adding marker or ink. Remember to study the leaves and stems carefully, looking for their unique characteristics. Slowly add value changes, texture, pattern, and shape to your drawing, and allow the leaves and stems to overlap. Notice that the plant lines are curved and flowing while the container and table lines are geometric.

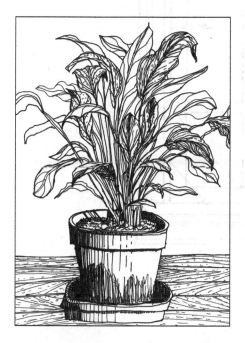

Leaves should have a floating quality.

Add value change indicating the front and back of the leaf.

Finally, finish adding value, texture, shade and shadow to the container, table and plant. For more guidance on how to incorporate these elements, refer back to previous chapters.

FABRIC WINDOW TREATMENTS

Window treatments exist in an interior space for multiple reasons - beauty, privacy, energy conservation and comfort. In your images, they provide an opportunity to add pattern and style to your interior drawings. It can be helpful to see window treatments in an elevation view before drawing them in perspective. The elevation view provides an opportunity to study the proportion and shape of the treatment style along with the pattern of the material.

In order to show window treatments realistically, it is important to draw the architectural features of the window correctly. Therefore, as you begin to draw an image of your room, be sure to take time to have an image of the type of window that is in your space.

This elevation of a double hung window shows the different parts of a window to include in your drawing:

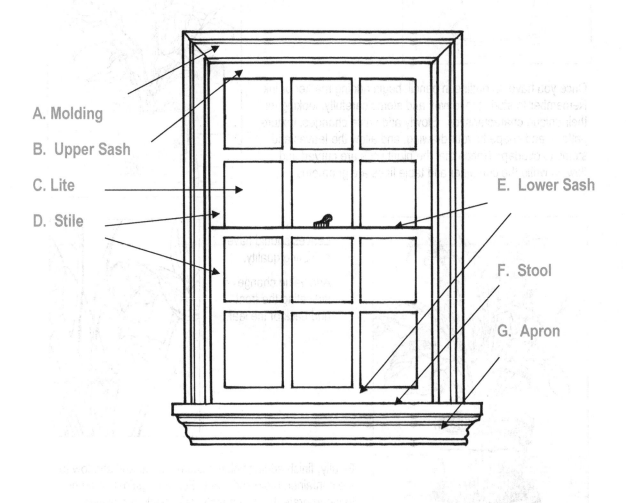

A. Molding

B. Upper Sash

C. Lite

D. Stile

E. Lower Sash

F. Stool

G. Apron

Chapter 7 - Interior Details

Top treatments can be installed inside the window frame. These types of treatment add softness and pattern to the window. The first is a balloon shade with fringe and the second is a roman shade with a decorative border.

The first two top treatments are shades designed to be outside the window frame. The last window has a cornice and a shade

Adding top treatment and draperies to a one-point perspective interior space begins by drawing box shapes. These shapes are then refined to create a valance or a rod and draperies. The images below illustrate a variety of window treatments. The following section will detail how to incorporate window treatments into your drawings.

WINDOW TREATMENTS IN ONE-POINT PERSPECTIVE

❶ From the vanishing point, draw a top guideline above the window frame to represent the top of the top treatment.

❷ Draw a second guideline below the window frame to represent the bottom of the top.

❸ Draw the left and right vertical sidelines representing the width of the top treatment. Divide this shape with the "X" into four sections.

❹ Draw side panels with horizontal lines on the left and right side of the window molding. Estimate the width of the panel.

❺ Next, place a tracing paper on top of this drawing and draw the type of window treatment and details. Alternatively, use light pencil on the drawing paper for this step.

These two window treatment drawings started with the preliminary drawing illustrated in the last page.

The first drawing depicts a cornice and side panels with rendered fabric.

❶ The center of the cornice was drawn with a fold.

❷ A cording trim line was added to the bottom of the cornice.

❸ The pattern is light valued horizontal and vertical plaid. It was added to the cornice and the side panels using the vanishing point as a guide for the horizontal pattern.

The second drawing is a rod with drapes without rendered fabric in the drapes.

❶ The rod was drawn in pencil using the top treatment as a guide from the preliminary drawing.

❷ The finals were added to the ends of each rod and their size is estimated.

❸ The rings were added in marker, first to the rod since they are viewed in front.

❹ The vertical drapes are drawn to imitate folds in the fabric.

DRAWING THE FIGURE

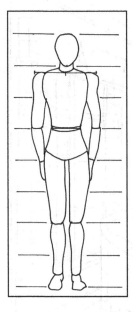

To add figures to your drawings, consider using photographs or entourage books as visual reference for generating ideas.

Furthermore, it is helpful to understand figure proportions before attempting to draw such images. I use a simple formula for deriving proportion that starts by first establishing the size of the head.

This image uses the artist figure as an example of how to derive proportions based on head height:

1 Total height equal to the 8 x the height of the head.

2 The shoulders are a bit more the 2 head heights.

3 The waist is about 1 head height

4 The legs are half of the body; 4 head heights

USING A TRACING FILE BOOK

The figures in the adjacent image were traced from a book called *Tracing File for Interior and Architectural Renderings*. The man in the foreground is eye level with the horizon line. The height of the two children was determined by their location in the image.

The figure on the left side has his eye level at 5'6" high. The other person is taller with a higher eye level. Both of these figures were drawn from photographs.

SKETCH PRACTICE PAGE

Adding accessories to your interior space will give interest and character to your drawings. In these sketches, the everyday items personalize the image and can often be the main interest of the image.

Helpful Hint

One way to become familiar with an interior space is to draw a floor plan. A quick way to do this is to use ¼ inch grid paper. This drawing started with the floor plan assisting with the location of the items.

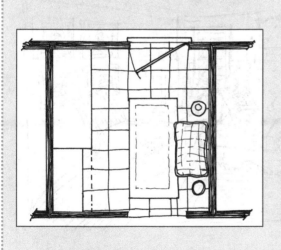

PUTTING IT TOGETHER

A successful drawing relies on both your drawing skills and your knowledge of the subject. A critical part of knowing your subject is to have images of the objects available for visual reference.   Visual references will

also expand your ideas as you tap into visual thinking. Images give us an opportunity to relate to our process in a non-linear way.

When drawing a realistic image of an interior space, your memory does not provide enough detail information. A picture of an item provides a visual reference to proportion, surface finish and decorative details. It also aids with generation of ideas for your image. You might think of this as an extra step that will slow you down and yet, having a visual reference will actually assist in finishing with a higher quality image.

I start a drawing project by searching for images or taking pictures of the accessories or items that I want to include in the drawing. I collect several images for each object. This gives me several options when I get ready to make the final choice. During the process, I often get started in one direction only to pause and turn in another.

I often put these images in a PowerPoint, organizing them by the different types of object. Next, I do a quick

sketch or two experimenting with the objects. Then, I choose the objects that I want to use for the drawing and put them into a document to make a black & white hard copy. This document becomes my visual reference while I am drawing.

Preliminary sketching of an interior space is as a creative strategy for generating ideas for the final design. Often, while sketching out an image, ideas for the composition will pop into my head. Moving my pen over the paper generates more ideas than just looking at pictures. I think there is something about being in motion that facilitates ideas popping into my head. Notice the shifts and changes from the preliminary sketch to the finished image that happened while I was sketching.

PRACTICE DRAWING QUICK AND SMALL

Consider practicing your hand drawing with quick small sketches. This is a great way to play around with drawing an image and avoid being bogged down by expectations of results.

Your expectations can frequently become an obstacle to developing your drawing skills. By using a small square to add a sketch, you can start and finish the drawing quickly and move on to another.

Here are a couple of examples of small, quick sketches, often called thumbnail sketches. These are 2" X 2" squares, just about the size of sticky note. Notice that I played around with different points of view

Three of these small images have a close up, cropped view that only shows a portion of the object.

The image of the lamp has a small portion of the decorative shade is seen and chair rail and wood paneling background was not actually there, it was added to create an interesting background.

The small plant has part of the leaves cropped off. The value on the container helps to add interest to the image

.
There is also a close up view of the mug with drawing supplies. The picture was added for a background.

One image includes all of a grouping with a view that is farther away.

Next time you have your sketchbook out, try drawing 2" X 2" squares and drawing quick, small images of interior objects.

Part III
Perspective Drawing

Perspective drawing in interior design provides a technique of depicting three-dimensional spaces on a flat surface and is a fundamental tool for communicating design ideas.

In this next section, we will focus on techniques for drawing one-point and two-point interior perspective drawings.

.

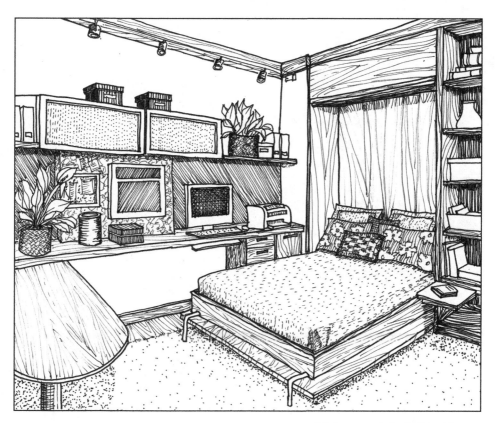

Interior space design by Erica Islas

Chapter 8

Interior One-Point Perspective Project

Perspective conveys the appearance of volume and spatial relationships on a flat surface. Understanding and using the grid method for drawing an interior space is the start to communicating your interior design solutions. In this chapter, you will draw a complete room in one-point perspective using a grid as your guide. For this activity, you will need a drafting surface, a straight edge tool, a triangle, and other drafting equipment.

GETTING STARTED

The goal for this chapter is to draw the interior of a one-point perspective room. You will begin by drawing a one-point perspective grid that will guide you as you draw the room, furniture and accessories. As you become more familiar with this process, you can adjust the grid as needed, but for now, use the suggested dimensions.

The grid is a visual tool to put the room in perspective. Each block on the grid represents 1 foot. This represents how we view a room with 1-foot blocks appearing to getting smaller as they recede into the room. For example, stand in the back of the classroom facing a wall. Look directly up or down at the 2-foot by 2-foot ceiling or floor blocks. As you look at these tiles, move your eye closer to the back wall and notice how they appear to get smaller and lose their square shape. Now, notice how the grid below shows the changes of the squares in perspective.

The steps to draw this grid are broken down into four sections: A, B, C, D. Using the one-point perspective grid sample below, notice these properties before you get started on your own grid:

① Each block represents 1 foot by 1 foot.

② The back wall with vertical and horizontal lines is an elevation with each block being a measurable size.

③ The floor, walls, and ceiling each have blocks that progressively get larger as they move away from the back wall.

④ Starting with the back wall, the lines on each surface extend from the corners of the back wall using the vanishing point. These are the perspective lines.

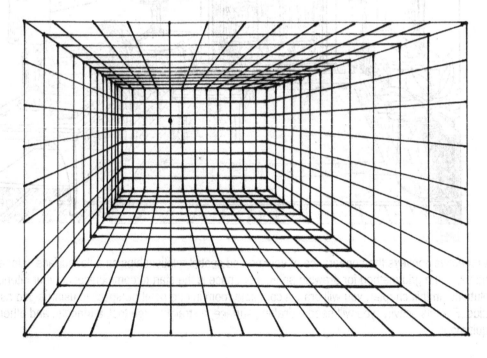

Chapter 8 – Interior One-Point Perspective Project

DRAWING A ONE-POINT PERSPECTIVE GRID

Tape a piece of tracing paper or drawing paper that is 24 inches by 30 inches to your drafting surface to draw the grid on. You will use this grid for the activity in this chapter as well as future projects.

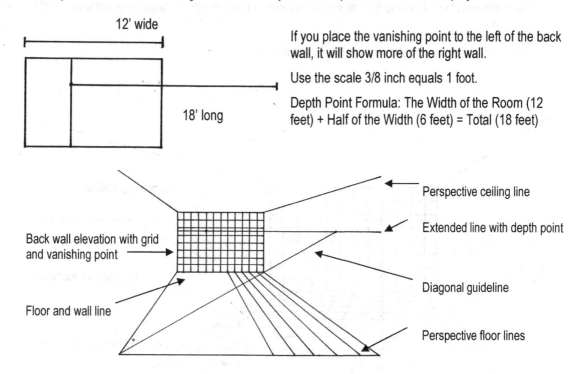

If you place the vanishing point to the left of the back wall, it will show more of the right wall.

Use the scale 3/8 inch equals 1 foot.

Depth Point Formula: The Width of the Room (12 feet) + Half of the Width (6 feet) = Total (18 feet)

A. THE BACK WALL AND DIAGONAL GUIDELINE

1. Draw a back wall as an elevation. Use the scale 3/8 inch equals 1 foot. Make the room 8 feet high by 12 feet wide.

2. Add a 1-foot-by-1-foot grid on the back wall using the 3/8 inch scale.

3. At 5 feet 6 inches high, draw a horizontal line that extends out from the wall. Put a vanishing point on the line 4 inches from the left side.

4. Draw the ceiling and floor lines starting from each corner using the vanishing point.

5. To find the depth point, mark a dot on the horizontal line from the vanishing point using the 3/8-inch scale that measures 18 feet.

6. To add the floor, draw perspective lines along the floor, starting with the floor/back wall boundary line, that extends from the vanishing point. (The image above shows five of these lines.)

7. From the depth point, draw a line with a straight edge that passes through the right bottom corner of the room into the floor area. This makes a long diagonal line that will be used as a guide for adding the flooring grid.

B. THE FLOOR GRID

❶ Finish drawing the perspective floor lines as shown in the image below.

❷ Draw horizontal lines along the floor at each point where the perspective lines intersect with the diagonal guideline.

❸ When the floor grid is complete, erase the diagonal guideline.

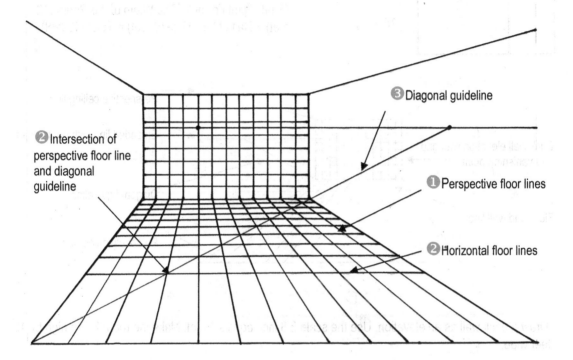

❸Diagonal guideline

❷Intersection of perspective floor line and diagonal guideline

❶Perspective floor lines

❷Horizontal floor lines

C. THE WALL GRID

1 Starting in the back, where the first horizontal line connects with the left floor line, draw a vertical line from the floor line to the ceiling line.

2 Continue this step to complete the left wall grid. These lines should be parallel to each other.

3 Repeat the above steps to create the right wall grid.

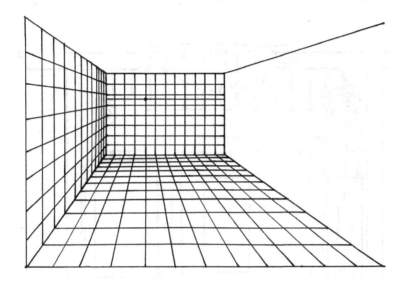

D. THE CEILING GRID

1 Add horizontal ceiling lines that connect the left and right walls.

2 Add perspective ceiling lines in the same manner as done to create the floor perspective lines.

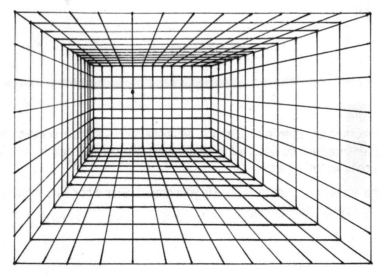

PRELIMINARY PLANNING

Before you begin using your grid to create a one-point perspective room, research your furniture selections, gather images of the furniture and record each object's dimensions. Then draw a scaled floor plan.

Below is a suggested simple floor plan for this project. The one architectural feature is the double door on the back wall. The floor plan shows the placement of the furniture that will be drawn on the grid later.

You are welcome to change any of the suggested architectural features or furniture for a different design solution.

Floor plans not to scale

DRAWING ARCHITECTURAL FEATURES

This page illustrates the doors on the back wall. Use your back wall grid when adding the door and the molding. Plan to have your door 6'-6"high and 6'-0" wide. Each door will be 3'-0" wide. The molding is aproximately 6" wide.

In the top drawing, note the follwing architectural features:

❶ On the left door, the vanishing point is in the center, thus you can see depth on both sides and the top and bottom.

❷ On the right door, you can only see depth on one side and the top and bottom.

❸ The doorknobs are drawn with the same steps used for furniture in Chapter 7.

❹ The molding around the doors extends out from the wall using the vanishing point as a guide for drawing the inside of the molding in the correct perspetive.

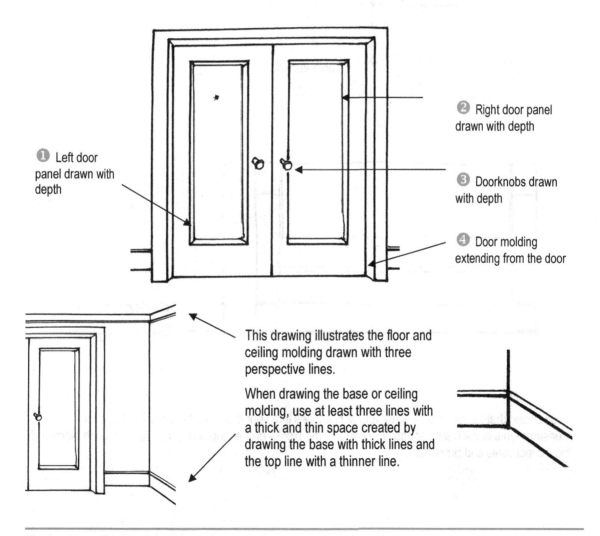

❶ Left door panel drawn with depth

❷ Right door panel drawn with depth

❸ Doorknobs drawn with depth

❹ Door molding extending from the door

This drawing illustrates the floor and ceiling molding drawn with three perspective lines.

When drawing the base or ceiling molding, use at least three lines with a thick and thin space created by drawing the base with thick lines and the top line with a thinner line.

SIZING OF FURNITURE

As you get ready to add furniture to your project, you will need each piece's measurements and features. Remember that often the best place to start is with a scaled elevation drawing of each piece of furniture. Use the scale ½ inches equals 1 foot. On this preliminary drawing, include the furniture dimensions.

Please note that in the interior design field exact measurements are very important. However, in a one-point perspective drawing, you may round these dimensions up to either 6 inches or 1 foot. For example, if the side table measures 21 inches high, you may draw it as 24 inches, or 2 feet. You will find that 6-inch or 1-foot dimensions are easier to use since the grid is set up in 1-foot blocks.

Use the drawings below as a reference for the features and dimensions of the table and sofa. Add dimensions to your own drawing as well.

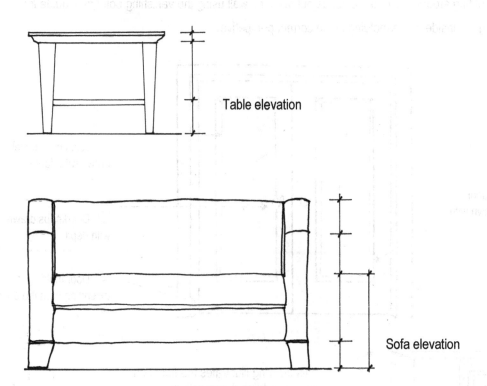

Table elevation

Sofa elevation

ADDING FURNITURE TO THE GRID

STARTING WITH THE BOX

When you begin to add furniture to the grid, start with a rectangular box shape. For example, see the four steps below.

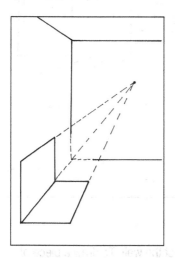 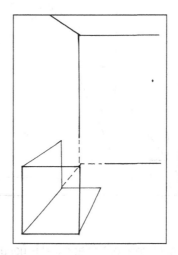 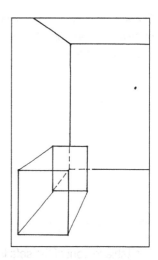

➊ Begin by drawing a box shape on the floor and the wall using your furniture measurements. You can use the grid and vanishing point as a guide.

➋ Add the horizontal and vertical lines at the front of the box. Each corner should be a 90-degree angle.

➌ Add the back horizontal and vertical lines. This corner should also create a 90-degree angle.

➍ To draw the front panel of the box, connect the two corners by using the vanishing points.

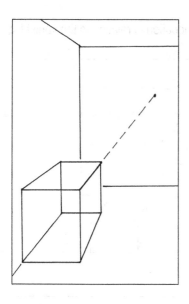

Helpful Hint

Draw your guidelines from the vanishing point in light pencil. You will end up erasing these guidelines after you finish drawing each furniture box. This technique keeps your drawings from having too many guidelines, which can get confusing as you add different furniture shapes

This also applies when drawing a flat-front box shape, which often incorporates over-extended lines. Erase any extra lines that extend outside the flat-front rectangle.

REFINING THE BOX INTO FURNITURE

Notice that in this illustration, each piece of furniture (couch, picture frame, carpet, and table) has a flat front.

The sofa shows the beginnings of being refined from a box shape to a finished sofa.

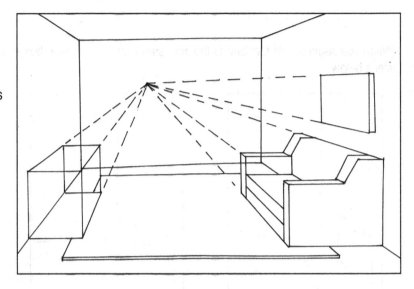

Box-Shaped Furniture in the Center of the Room

The table in front of the sofa is the only piece of furniture that is not against the wall. To draw a piece of furniture in the center of a room, follow these steps:

❶ Draw the box shape on the floor grid at the location in the room that you want the table to be.

❷ Use the right wall grid as a reference to determine the height of the table. Note the dotted guidelines. They provide the flat front of the table

❸ To draw the side lines of the tabletop, connect each corner of the flat front to the perspective guidelines that originate from the vanishing point.

❹ Notice that you can only see the right side of this table. The table's location in relation to the vanishing point determines what sides of the piece are visible.

Helpful Hint:

While working on perspective drawings, remember to use other chapters in the book as a reference for drawing techniques.

For example, refer to Chapter 7 for a refresher on how to draw the plant and planter.

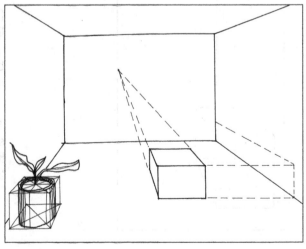

FINISHED LINE DRAWING

Notice the following composition elements in this drawing:

1 The overlapping plant, table, and tree create depth on the left side of the drawing.

2 The cropped ceiling helps to avoid having too much negative space at the top of the image.

3 The plant and table leading off the page adds interest to the two bottom corners.

4 Placing the vanishing point to the left, instead of the center, avoids a bull's-eye. A bull's-eye in the center of an image can distract from the rest of the image.

5 There is an overall informal balance between the pieces of furniture on each side of the room.

Helpful Hint

Before rendering your perspective line drawing, taking time to render a floor plan can be an important step. It provides you with an opportunity to practice, think through where you will put your textures and patterns and be uses as a visual reference when you are working on the rendering.

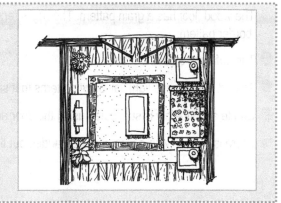

FINISHED RENDERED DRAWING

Above is a finished rendered drawing of a one-point perspective room reviewed in this chapter. Notice these details in the rendered version:

❶ On the left, the plant container has texture which is lighter in the front and gradates to darker has the texture is drawn close to the wall.

❷ The wood floor has a grain pattern. The rug is rendered with stippling and has a geometric shaped border pattern.

❸ The sofa has a floral pattern.

❹ The art pieces have light values of patterns that suggest shapes.

❺ Shade and shadow is suggested using the door as the light source.

❻ There is a porch with a plant that was added out the door windows.

Chapter 8 – Interior One-Point Perspective Project

Creative Strategy
TAKING A BREAK

As you move on to more complicated drawings, you will find yourself spending more time working on them. Taking a break is a critical part of the drawing process. When you take breaks, it allows you to recharge your focus and enthusiasm as well as providing you with a refreshed look at a piece each time you sit down to work on it.

For example, I have discovered that I work best one or two hours at a time. I stay relaxed, and I am excited about the progress and enjoy myself. However, if I work on a drawing longer than this, I start to get discouraged, and I become more critical of my work, and I get distracted.

Breaks do not have to be long. They can be simple like taking a walk or making a snack. Or, you can put the drawing aside until another day. Taking a break becomes a recharger. You likely will come back to the drawing refreshed.

With a refreshed perspective, you are then able to see your previous work much more clearly. You will be able to see needed adjustments that you may not have seen before, or you may see that a drawing is at a strong completion point, and thus avoid over working it.

Overall, breaks are essential to creating successful drawings

Chapter 9

Interior Two-Point Perspective Project

In this chapter, you will draw a room in two-point perspective using a two-point perspective grid as your guide. You will need a drafting surface, a straight edge tool, a triangle, and other drafting equipment for this project.

GETTING STARTED

The goal for this chapter is to draw the interior of a two-point perspective room. You will begin by drawing a two-point perspective grid that will guide you as you draw the room, furniture, and accessories. As you become more familiar with this process, you can adjust the grid as needed, but for now, use the suggested dimensions.

Each block on the grid represents 1 foot and shows change in perspective, just as you saw on the one-point perspective grid.

PRELIMINARY PLANNING

Before you begin using your grid to create a two-point perspective room, you will need a scaled floor plan, images of the furniture and accessories you plan to use, and their measurements.

Below is a suggested floor plan to demonstrate the project. It is a simple bedroom with a hallway behind the bed. The perspective is drawn from the viewpoint of a person standing by the corner of the bed. You are welcome to shift or change any of the suggested architectural features or furniture for a different design solution in your own project.

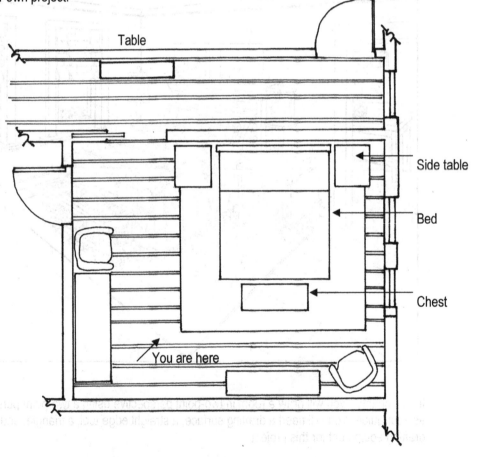

DRAWING A TWO-POINT PERSPECTIVE GRID

The steps to draw this grid are broken down into five sections: A, B, C, D, E. Using the two-point perspective grid sample below, notice these properties before you get started on your own grid:

❶ Each block represents 1 foot by 1 foot.

❷ The perspective lines on each wall originate in the back corner.

❸ The floor, walls, and ceiling each have blocks that progressively get larger as they move away from the back wall.

❹ The horizon lines originate from the left and right vanishing points.

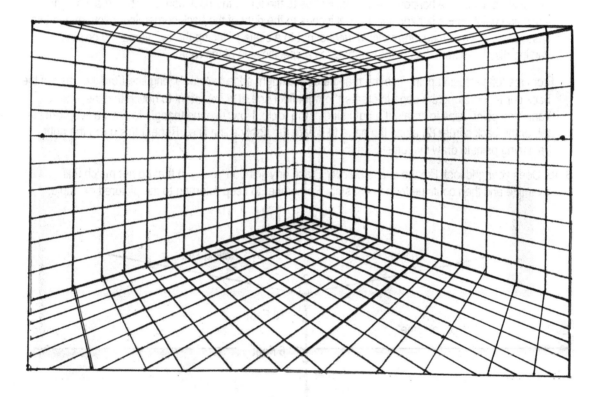

A. THE BACK CORNER AND VANISHING POINTS

Tape a piece of 24-inch by 36-inch tracing paper or drawing paper onto your drafting surface to draw your two-point perspective grid. Once the grid is complete, you will be able to use it for this project as well as others in the future.

❶ Start with a piece of plain 24-inch by 36-inch paper.

❷ Draw a vertical line in the center of the page. This will be the back corner. Using your architectural scale, mark the back corner line every 3/8 of an inch from 0 to 10 following the scale: 3/8 inch equals 1 foot. These marks should end at the ceiling, which is 10 feet high.

❸ Mark the bottom corner with a (1) and the top with a (2). We will use these points to create the floor lines and ceiling lines.

❹ Draw the horizon line through the 6-foot mark. Next, using a ruler or an architectural scale, draw a left vanishing point on the horizon line 10 real inches to the left of the back corner. Then draw a right vanishing point on the horizon line 14 real inches to the right of the back corner line. (Note that in this case, these are *real measurements in inches*, not to a scale.) Placing the two vanishing points at varying distances from the center emphasizes a specific side of the room.

❺ From the left vanishing point, use the triangle as a straight edge to draw a long line from the base of the back corner (1) along the page to create the right floor line. It is important to notice that the left vanishing point helps to create the right side of the room. Now, use the triangle to draw a line from the top of the back corner (2) along the page to create the right ceiling line. Repeat this using the right vanishing point to draw the left side of the room.

Stop! Does your grid look like the image below? Erase any lines on your grid that do not match this example and then continue to the next step. Remember to keep referring to the pictured examples.

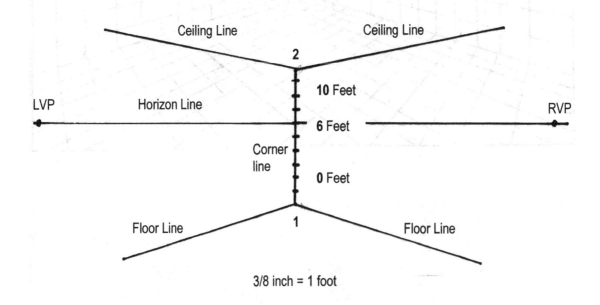

B. THE WALL GRIDS

1 Extend the back corner line downward by 10 feet starting from the corner point (1) to create a guideline for the floor grid. Mark the end of this line as point (3).

2 Using your straight edge to align the right vanishing point (RVP) with point (3), mark point (4) where this line crosses the floor line. Draw a line from point (3) to point (4). Now, use the triangle to draw a vertical line from the floor line at (4) to the ceiling line. Mark this as point (5). This line represents the edge of the right wall. Repeat the same on the left side by using the left vanishing point (LVP). Mark a (6) where this line meets the floor line. Draw a vertical line from point (6) to the ceiling line and mark a (7). This creates the edge of the left wall. Now, erase any extra guidelines that you may have.

3 Next, draw a diagonal line from point (5) to the base of the back corner (1). Repeat this on the left side from the (7) to the (1).

4 Then add horizontal wall lines by placing one point of your triangle on the left vanishing point and the other point on the first scale mark at the base of the corner line (1). Put your pencil on the scale mark on the back corner line and draw horizontal lines out from each scale mark along the right wall. Repeat the same steps using the right vanishing point to create horizontal lines on the left wall.

5 To add vertical wall lines to the right wall, draw a vertical line from the floor to the ceiling at each point where the diagonal line (5 to 1) crosses a horizontal wall line. Continue to draw these lines until you reach the (4)-(5) line. Once you are finished, erase the diagonal line.

6 Repeat the above steps to add vertical lines to the left wall by drawing lines from the floor to the ceiling at each point where the diagonal line (7 to 1) crosses a horizontal line. Continue to draw these lines until you reach the (6-7) line. Once you are finished, erase the diagonal line.

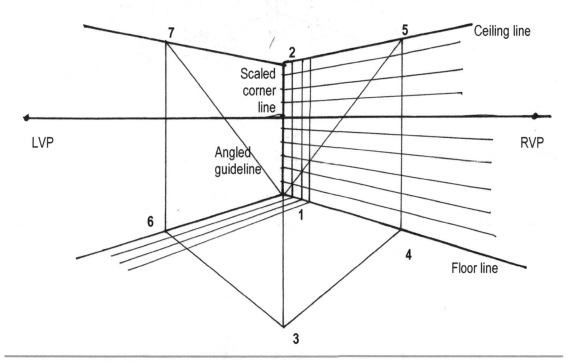

C. THE FLOOR GRID

❶ The lines on the floor are simply lines that are parallel to the initial floor line. Begin each line at the points where the vertical lines meet the floor line. Extend these lines beyond the (3)-(6) line.

❷ Continue to add floor lines by starting each line at the points where the vertical lines on the left wall meet the floor line. Extend these lines beyond the (3)-(4) line.

❸ Now, erase the diagonal guidelines on the walls and the extend (1)-(3) line from the floor area.

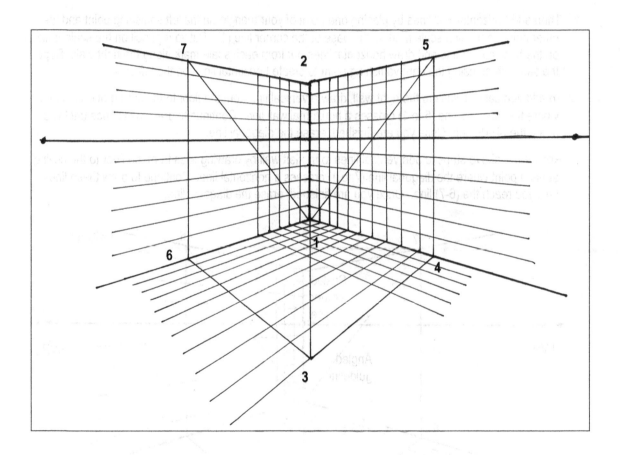

D. EXTENDING THE WALL

① From the base of the back corner point (1), count five blocks along the floor line and mark it point (8). This is half of the 10 blocks.

② From point (8), count another five blocks and mark this point (9).

③ From point (9), count up five blocks along the wall and mark this point (10).

④ Using a straight edge, draw a line through points (8) and (10) and up to the top line. Mark this as (11).

⑤ When this diagonal line crosses a horizontal line, draw a vertical line from the ceiling line to the floor line. As you add vertical lines, they will progressively get farther apart, creating the perspective of the grid.

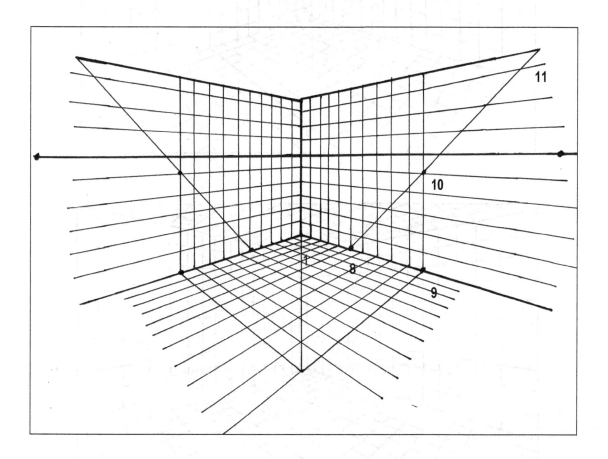

E. FINISHING THE WALL GRID

1 Repeat the steps in section D on the other wall. Once you are done, erase the diagonal guidelines.

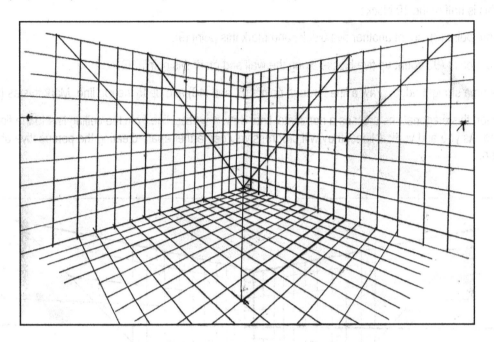

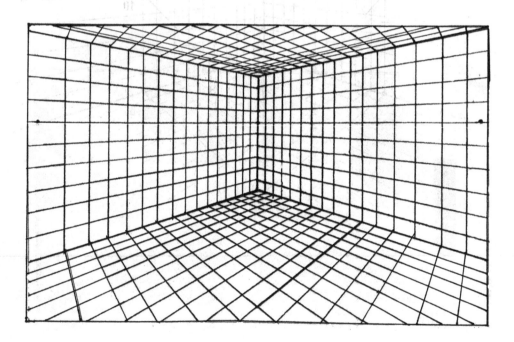

A finished two-point perspective grid

ADDING ARCHITECTURAL FEATURES

WINDOW AND ROMAN SHADE

To add a window and Roman shade to your grid, follow these steps:

1 Draw the window molding approximately 6 scaled inches wide. Use the grid as a guide to measure this width and to find a place for the window along the wall.

2 Add the inside and left corner of the molding by estimating their depths. Use a perspective line to create the corner.

3 Add the windowsill by visually estimating the depth and width of it.

4 Add the inside of the window by visually estimating its width.

5 Add the Roman shade on the inside of the window molding. Visually estimate the width, and use the left vanishing point to create the horizontal lines of the shade while adding a slight curve to the sides of the shade.

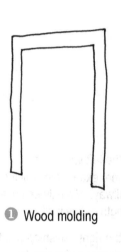

1 Wood molding

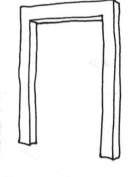

2 Window molding

3 Windowsill

4 Wood molding

5 Roman shade

Helpful Hint:

When drawing architectural features in perspective, find an example that you can look at as you are drawing. In this case, it can either be an actual window with molding or an image of one. Being able to see an example will help you understand the depth of the feature so you are able to draw it in perspective.

CEILING, DOOR AND HALLWAY

To add the ceiling and door to your grid, follow these steps:

❶ Draw the ceiling molding with three perspective lines.

❷ Draw the door molding approximately 6 scaled inches wide. Use the grid as a guide for this measurement.

❸ Add the inside and right corner of the door by visually estimating the width. Use an angled perspective line to create the corner.

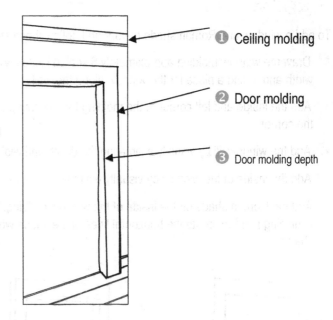

❶ Ceiling molding

❷ Door molding

❸ Door molding depth

To add the hallway to your grid, follow these steps:

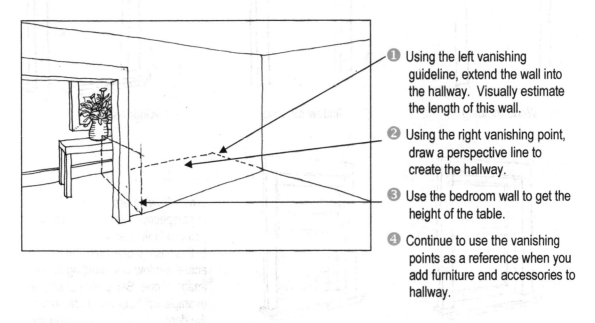

❶ Using the left vanishing guideline, extend the wall into the hallway. Visually estimate the length of this wall.

❷ Using the right vanishing point, draw a perspective line to create the hallway.

❸ Use the bedroom wall to get the height of the table.

❹ Continue to use the vanishing points as a reference when you add furniture and accessories to hallway.

SIZING OF FURNITURE

Before you begin adding furniture, you need to record the measurements of each piece of furniture. Do so by drawing an elevation of each piece using the scale: ½ inch equals 1 foot. Mark the furniture's scaled measurements. Notice in the elevations below that the scaled dimensions are noted to the side of the drawing.

In the interior design field, exact measurements are very important, particularly for elevation drawings. However, in a two-point perspective drawing, you may round these dimensions up to either 6 inches or 1 foot. For example, if the headboard measures 55 inches high, you could round that up to 60 inches, or 5 feet. You will find that 6-inch or 1-foot dimensions are easier to use since the grid is set up into 1-foot blocks.

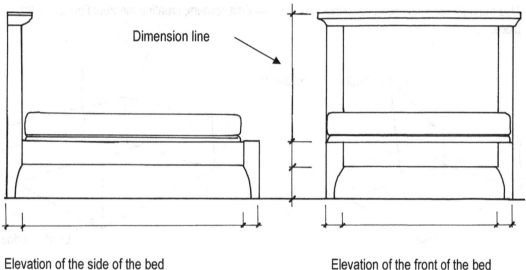

Dimension line

Elevation of the side of the bed   Elevation of the front of the bed

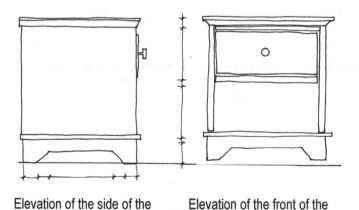

Elevation of the side of the
table

Elevation of the front of the
table

ADDING FURNITURE TO THE GRID

Once you have your scaled elevation drawings, follow these steps to add furniture to the perspective grid:

① Start by drawing a box shape with the width and depth of the box on the floor of the grid and the height and width of the box on the wall of the grid.

② Use the left vanishing point and the top left corner point to create the left edge of the piece of furniture. Extend this line out just above the front corner of the bed.

③ Draw a vertical line from the front left corner of the bed to the floor. This is the leading edge of the piece. It must cross the horizontal line from step 2, which is the correct height of the bed.

④ Use the back, right corner and the left vanishing point to make the right edge of the piece of furniture.

⑤ Use the right vanishing point to connect the right and left corners, creating the front line on the top of the piece.

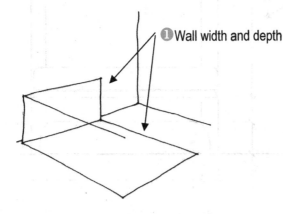

① Wall width and depth

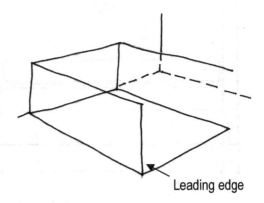

Leading edge

② Left side beginning

③ Left side complete

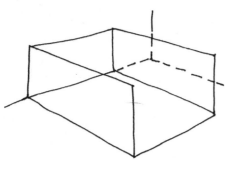

④ Right side

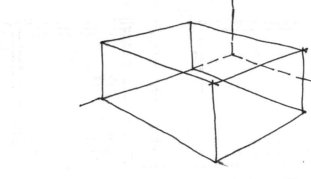

⑤ Front side

Chapter 9 – Interior Two-Point Perspective Project

FURNITURE GROUPINGS

As you add more pieces of furniture to your drawing, notice how the furniture is grouped in the examples below. The first illustration includes the bed, headboard, side tables, and bench at the foot of the bed. The dotted lines represent parts of the furniture that are hidden from view.

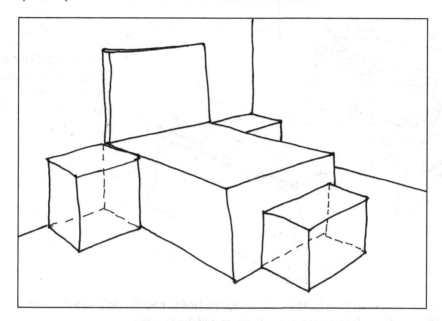

Now, see how this furniture grouping is included in the drawing of the room.

ADDING DEPTH AND RENDERING

Here is an example of a finished bed with a headboard, a quilt, sheets, and pillows.

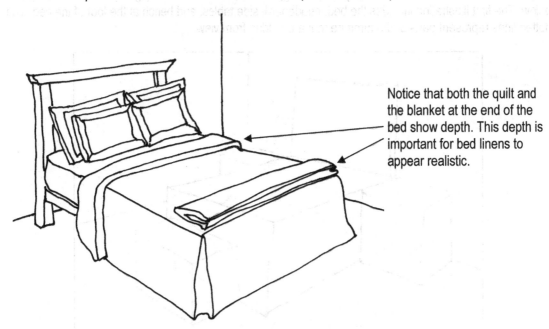

Notice that both the quilt and the blanket at the end of the bed show depth. This depth is important for bed linens to appear realistic.

Below is a rendered image of the furniture shown above. Notice how the right vanishing point helps to keep the pattern on the quilt in perspective with the rest of the drawing.

FINISHED LINE DRAWING

Above is a finished line drawing of the two-point perspective room reviewed in this chapter. You may add your own flavor to this project as well. Think about adding things, such as a container plant, various accessories on the side tables, different window treatments, a view outside the window, or any other unique features.

FINISHED LINE DRAWING

Above is a finished line drawing of the two-point perspective room reviewed in this chapter. You may add your own flavor to this project as well. Think about adding things, such as a container plant, various accessories on the side tables, different window treatments, a view outside the window, or any other unique features.

Creativity Strategy
THE DRAWING COACH

It is easy to have doubts about learning new skills and the progress you are making. However, even if you are nervous about showing your work to someone else, I have found that it can be very helpful to have a supportive person you can rely on to answer questions and to provide constructive feedback. You find this when you are a part of a community of artists or when you join a class.

For me, as I moved from the interior design into visual arts, I began learning new skills. I went to drawing classes and noticed that I was very sensitive about my work. First, my work was not very good. Second, I did not want anyone to be critical of my work. Nevertheless, I knew feedback was critical for staying on track. I looked forward to learning more about what I was doing well and what steps I needed to take to move forward. With this in mind, I realized that I needed a drawing coach. Fortunately, at that time a coworker, Ellen, agreed to help me.

As my drawing coach, Ellen was just what I needed to continue progressing. I felt comfortable showing her my work and, I would ask her questions to get her impressions on the piece.

I recommend that you find a drawing coach, too.

Interior space designed by Connie Riik

Chapter 10

Advanced Perspective Techniques

Now that you have a good understanding of drawing in perspective, we can incorporate methods to simplify these drawing steps. This chapter will explore one and two-point perspective interiors using several quicker set up methods.

This includes using the back wall, starting with a single vanishing point, a simplified grid technique and using a photograph by added grid.

GETTING STARTED

In this chapter, we will explore a variety of tips and techniques for creating more advanced perspective images. However, before we begin, it is important to review a few basic strategies for getting started.

Here are several initial tips you can use when preparing a perspective drawing:

① Start by thinking about and planning your drawing in your sketchbook. This is a place you can feel free to develop ideas for the image. It provides a space to identify the furniture pieces and objects to include in the finished image. It is helpful to keep this sketch in view when working on the image. You can add to the sketch or rework ideas on the same sketch page.

② Draw a sketch of the floor plan of your design. This will provide spatial relationships of space and furniture and is an important visual reference while working on the finished image.

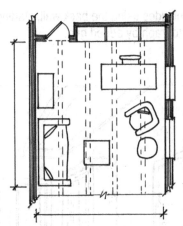

③ Set up your perspective drawing onto a backing. I use a 9" X 12" cardboard piece for the backing that comes from the back of a drawing paper pad. The 8 1/2" X 11" drawing paper is then taped to the cardboard backing to provide support for using the T-square and triangle while drawing the image.

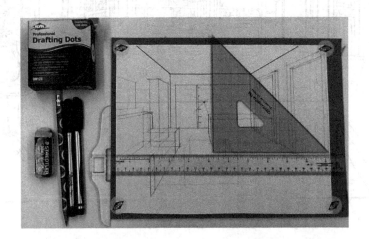

USING THE BACK WALL

In a one-point perspective, the back wall is in a measured scaled grid. This provides an opportunity to extend a built in wall cabinet using this back wall grid.

In this series of drawings, the goal is to design a built in cabinet with a small sleeping space, study and storage area. There is an imaginary vanishing point to the far left of the cabinet, which provides the reference for angled horizontal lines.

Floor Plan Drawing

This floor plan was drawn on grid at ¼" scale and provides a visual reference for the space.

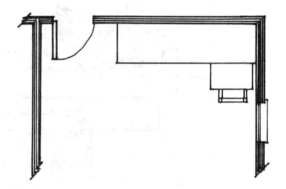

Preliminary Planning Sketch

The preliminary planning sketch shows the design ideas for the cabinet and provides time to work through design ideas.

Several planned features included shelves facing the left side of the cabinet, the bed placement and the desk that extends out from the cabinet on the right.

STARTING FROM A SINGLE VANISHING POINT

Drawing and sketching small groupings of furniture can start with a single vanishing point. This furniture grouping is a quick practice sketch done using this method. The key to this technique is to draw the foreground piece of furniture first. For example, in the image above, the trunk is in the foreground and it was the first shape added to the drawing. The trunk shape and size becomes a guide to determine the size of the other pieces. Each subsequent object is drawn in relationship to the trunk size. This sequence is outlined in the drawings on the following pages.

The fabric designs used in the pillow are shown in the shapes below. An effort was made to select a variety of different patterns. Starting on the left, these patterns include a paisley, a small flower, diagonal, crisscross, zig-zag and a square design.

Here are the steps taken to draw the furniture grouping.

Start With Foreground Box

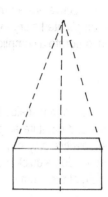

❶ Draw the rectangular flat front of the chest in the center and bottom of your 9" X 12" drawing paper. Remember that all the other pieces of furniture will be drawn around this shape.

❷ Find the center of the rectangular shape and draw a guideline up to a vanishing point at the top of your drawing paper. This center guideline will be used when drawing the other objects.

❸ Use the vanishing point to draw the sides of the top.

❹ Connect the two sides with a line to complete the back of the trunk.

Add Sofa Base

❶ Draw the two sides of the sofa equal distances apart using the center guideline. Draw the sofa wider than the trunk and locate the base of the sofa slightly above the truck.

❷ Use the perspective guidelines to draw the depth of the arms.

❸ Draw horizontal lines indicating the sofa base and seat cushions estimating the width.

❹ Use the perspective guidelines to draw the sides of the cushions.

❺ Intersect the two side seat lines with a horizontal line.

Add Remaining Elements

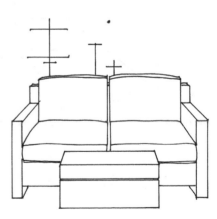

❶ Finish drawing the sides and back of the arms.

❷ Divide the seat cushion in half using two lines that are from the vanishing point.

❸ Draw the back cushion and a sofa back on each side above the arms.

❹ Draw guidelines for the lamp and two container shapes.

MULTIPLE PERSPECTIVES

This perspective project utilizes an abbreviated one-point perspective grid technique to create a space with more complicated elements. For this project, we will place one and two-point perspective furniture in the grid, incorporate a complicated bookshelf, an open door and ceiling beams.

Create Scaled Floor Plan.

Again, start with a scaled floor plan showing the architectural features and furniture as a visual reference. Here the floor plan for this perspective project was drawn on ¼" scale paper. The grid paper provides a quick method for drawing the floor plan to scale. The width of the room is 12'-0" and the length is 16'-0".

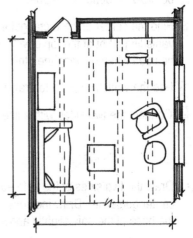

Create Quick Grid

Using the techniques for the One-Point Perspective Project in Chapter 8, create an abbreviated one-point perspective grid with the horizon line lowered to 4'-0" off the floor line. For this project, we will only be creating the back wall and floor grids.

Start with a back wall in measured scale, establish a horizon line, place your diagonal guideline then build your back wall grid and floor grid.

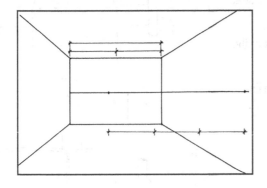

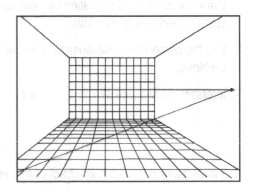

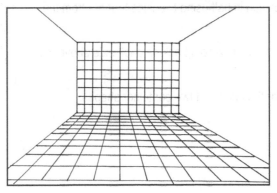

Final grid

Sketch Outline of Interior Space

Place a tracing paper overlay on the grid and add sketched furniture shape blocks to the floor.

These shapes are drawn into blocks using the method outlined in chapter 8. Architectural features such as the door, bookcase and windows are blocked out.

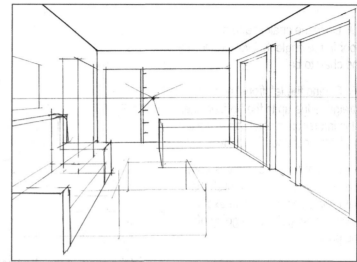

Refine the Bookcase

Start by drawing a grid for the bookcase shelving and cabinet fronts. Use the original vanishing point to establish the inside depth of each shelf as depicted in the second image.

Add the Open Door

Add a sketch of the top and bottom lines for the open door.

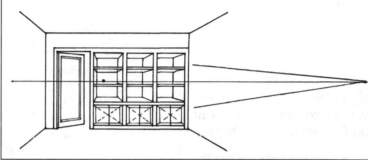

Next, use these lines to create an alternate vanishing point on the horizon line. Extend the top and bottom lines of the door until the both intersect with the horizon line as shown in the image above. This new vanishing point will be used to refine the door shape and the molding on the door.

Helpful Hint

Keep in mind, while there will be only one horizon line in your image, there may be multiple vanishing points on the horizon line.

The horizon line represents the eye level of the viewer in the interior space. The furniture in the picture maybe placed in the room at different angles. Therefore, these furniture pieces will have different vanishing points on the horizon line. Some may be in one-point perspective, while others may be in two-point perspective.

Add Two-Point Perspective Chair

1 Draw the chair shape on the floor in the angle that you would like the chair to be.

2 Extend the left front line with a straight edge up to the horizon line. This intersection establishes the left vanishing point for the chair.

3 Repeat the second steps from the right front line to establish the right vanishing point. The example shows these guidelines going off the page.

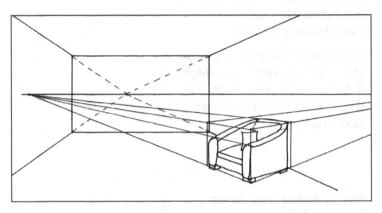

4 Note the side table next to the chair used the same steps. You can see an example of this in the next drawing.

Add Ceiling Beam Feature

1 Starting on the top left side, using the grid, count 2'-6" and mark this as the left side of the first beam. Mark 6" for the width of the beam.

2 Continue this pattern across the ceiling.

3 Using the original vanishing point, draw the sides of the ceiling beams.

4 Add a 3" beam on either side of the room.

5 The position of the beams to the vanishing point will determine how much each beam is shown.

Finished Line Drawing

Finished Rendered Drawing

This drawing was rendered with light valued textures and patterns. The goal of the rendering was to create a room that was easily viewed by highlight the furniture and objects with the rendering.

USING A SQUARE GRID FORMULA

This is another technique for creating a one-point perspective grid that can be used for quick sketches. This technique uses a back wall that is drawn in a measured scale square. This drawing series of an art gallery demonstrates the technique. It starts with a back wall that is 10" X 10", and shows several new perspective drawing features including the addition of track lighting, a back room and a figure.

Define the Back Wall

Draw a square in a measured scale that is 10' X 10'. Mark each foot division on the box shape

Establish Vanishing Point

Establish the horizon line and vanishing point. In this drawing the horizon line is 5' high and the vanishing point is 4' from the right side. This will give the left wall more of a focal point.

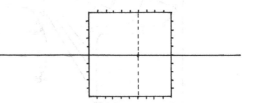

Grid Front Square and Wall Divisions

① Starting from each corner, draw the wall and ceiling lines using perspective guidelines from the vanishing point.

② Establish the length of the room by drawing a square where you want the room to end.

③ Draw the horizontal lines on the left wall every two feet using the vanishing point and perspective guidelines.

④ Using a straight edge, draw a diagonal line from the bottom left corner to the top left corner. This will divide this wall in four equal parts.

⑤ Draw vertical lines where the diagonal and horizontal lines intersect. These will be referred to as wall guidelines in next set of instructions.

⑥ To divide the floor, where the vertical lines hit the floor line, draw horizontal lines. These lines extend up the right wall as vertical lines. The figure in the right back corner has their eye level the same as the horizon line.

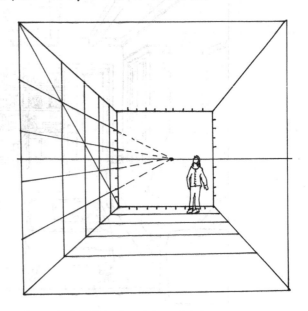

Chapter 10 - Advanced Perspective Techniques

Add Art Images to the Walls

① Use a tracing paper overlay to start adding features.

② Start by establishing the width of the first frame on the left wall. Use that first wall guideline as the right side of the frame and estimate the width.

③ Establish the height of the first frame using the vanishing point to define the angles. The middle of the frame in this image is 5'-6", which is eye level.

④ Continue to add frames to the left side using the front frame and the wall guidelines.

⑤ Add lines to the right wall using the floor lines as a guide.

⑥ Repeat the process as above to add picture frames on the right wall.

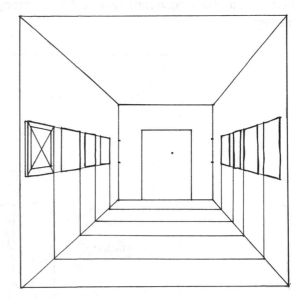

Add Additional Features

① Draw track lighting to the ceiling by extending the wall guidelines on the ceiling plane. Use the right wall to measure the depth.

② Draw the bench and two stands by using the wall guidelines on the floor plane. The height of the bench was measured from the right wall.

③ Draw a back room floor line estimating the length of the back room. Add frames estimating their size.

④ The figure in front was copied from a photograph. The figure height is established from the 5'-0" horizon line and a perspective guideline on the floor as depicted in the image.

⑤ Two small figures were sketched free hand and their sizes were established by using the horizon line and perspective guideline.

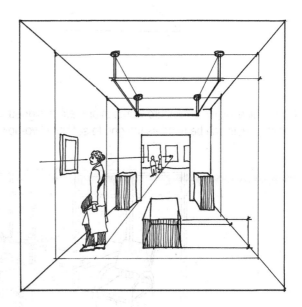

Final Rendered Drawing

Here is the final drawing. A small amount of line rendering was added to highlight the key elements of the drawing.

STARTING FROM A SKETCH

Open office systems furniture workstations are designed in a grid format using a support panel system. These panels can be used as the grid to set up a two-point perspective work place image.

Create Preliminary Planning Sketch

This drawing series of a simple open office workstation started with the preliminary planning sketch shown on the right.

A horizon line was established at a 5-foot mark on the right side. This became the proportion gauge for the height of the furniture pieces. The vanishing points were imagined far off the page.

The sketch includes features such as overhead cabinets, the counter top, file cabinets, and chairs.

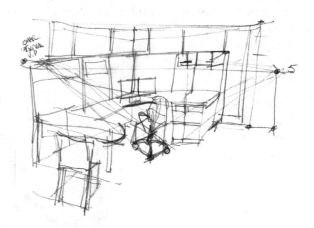

Add Panels & Guidelines

This drawing demonstrates how the chairs were refined with two vanishing points on the horizon line that are different from the vanishing points used to draw the system panels.

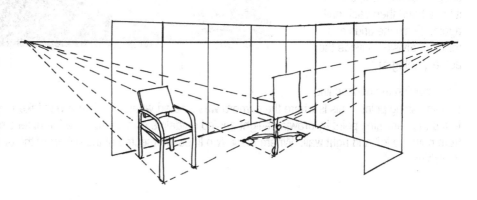

Continue Adding Details

This is another preliminary drawing with more details and refining on the furniture pieces.

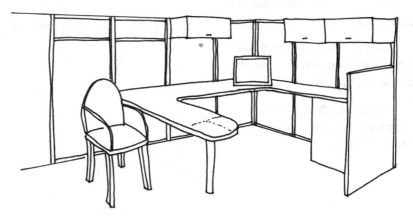

INSPIRATION FROM A PHOTOGRAPH

Photographs can provide ideas and inspiration for your interior drawings. Plan to use photographs you have taken unless you have permission from the photographer or the designer to use their image. I have successfully contacted photographers and interior designers using their website email addresses.

The photograph of this bedroom is of a room designed by Connie Riik.

Setting Up the Photograph:

① Divide the photograph shape into quarters, dividing it in half and then each half again. Draw the grid lines for each division outside the actual photograph.

② Establish the horizon line and vanishing point. This is where the camera was located when the picture was taken. In this image I found the vanishing point by drawing a light angled guideline with a straight edge extending the ceiling lines from both the left and right wall. Where these two lines intersected at the center of the door, became the vanishing point.

Starting Your Drawing:

① Draw a rectangular shape that defines your image boundaries.

② Add the proportion grid used in the photograph.

③ Using the photograph as a guide, estimate location of the horizon line and vanishing point.

④ Add furniture blocks following the grid pattern from the photograph and the vanishing point.

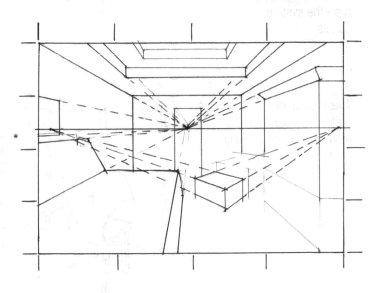

Finished Line Drawing

On a tracing page over lay of the first drawing, refine the furniture blocks. Add the plants, pillows and other accessories to your image to complete a line drawing.

Rendered Line Drawing

Add pattern and texture to the objects. Keep in mind how the variety of value helps distinguish layered objects.

Notice on the bed that the throw blanket is a darker value than the bedspread pattern.

Interior space design by Connie Riik

Another Example

Here is another example of using a photograph as a visual reference for drawing architectural features and details. The grid added to the image assists with the proportions of the door.

DRAWING OUTSIDE

Urban sketching is an active global drawing focus made up of local artists practicing drawing on location. Journalist Gabriel Campanario started the Urban Sketchers movement on Flickr in 2007. I enjoy this forum as a means for drawing practice and art expression.

The city I live in has wonderful buildings that are great inspirations for drawings. In these three examples, a photograph provides a visual reference for drawing the proportions and architectural details.

The Florida Theatre, with a rich history as a venue for music and art, is one of many historic buildings in downtown Jacksonville. The building has many decorative architectural details that I enjoyed including in the drawing.

In this sketch of downtown Jacksonville, I was sitting across the river on a park bench. On site, I sketched the outline of building structures, the landscape and the river scene. This sketch and a photograph provided the visual references for me to finish the detailed drawing later.

This drawing is one of a series of historic fire stations still used in Jacksonville, Florida. The #10 station is located in the historic Riverside district.

PUTTING IT TOGETHER

Traveling can provide a great opportunity for drawing new subjects. Plan to pack a small case with your drawing supplies and a sketchpad. Often there is time as you are sitting and waiting to pull out your sketchbook and drawing makers. I look forward to finding opportunities to sketch and work on drawings. Here are two drawings done while I was traveling.

The first line drawing I completed while traveling back to Florida from Austria on the plane. I was captivated with this image on the cover of a brochure that I had in my travel bag. It was taken from the top of a building in Salzburg. I used it for my inspiration and it was easy to work with while sitting in the tiny airplane seat. Luigi Caputo, who I contacted to get approval for using his image, took the photograph. He has a wonderful website with other exciting photographs.

This image came from a trip to Vancouver and was the view out the hotel window. I pulled up a chair to the window and used my small travel sketchbook. I had just enough time to draw an outline of the scene. As I traveled during that day, I added the architectural details and trees to the drawing.

Creativity Strategy
LIFELONG LEARNING

We bring all of our lifelong experiences to the process of learning. As we progress through life, we learn more about ourselves, and in turn can support ourselves when we recognize the ways we like to learn, the importance of practicing and the willingness to experience new concepts.

For students who have just left the confines of high school, it is important to realize that just because you did not learn a skill growing up does not mean you cannot learn it now. Forget what others may have told you in the past about your art skills and focus on living life with the passion to learn new concepts, no matter what others say.

For older students, it is not too late for you either. Do not use your age as an excuse; there is just as much benefit in beginning to learn how to draw as a young college student as there is in learning as an older student. For younger students you have the benefit of years ahead of you to practice and hone your skills. For older students, you have the benefit of a life full of experiences and a clearer understanding of how you learn best to help give you a strong start.

There is a tendency to explore and be more adventurous as we get older. As I have reached my 50's, I am finding it easier to ignore the critical opinions of people from the past. The younger you can realize this, the earlier your creativity will open up and expand.

So, not matter what stage of life you are in, remember that learning is a lifelong process that draws on all of your experiences. There is always an opportunity to embrace the excitement of learning new skills

Chapter 10 - Advanced Perspective Techniques

Bibliography

Brainard, Shirl *A Design Manual* New Jersey: Pearson Hall 2006

Campanario, Gabriel, *The Art of Urban Sketching, Drawing on Location Around the World,*

Ching, Francis D.K. *Design Drawing.* New York: Van Nostrand Reinhold, 1998

Ching, Francis D.K. *Drawing, A Creative Process.* New York: Van Nostrand Reinhold, 1990

Hanks, Kurt *Rapid Viz, A New Method for the Rapid Visualization of Ideas.* California: William Kaufmann, Inc., 1980

Koenig, Peter A. *Design Graphics, Drawing Techniques for Design Professionals.* New Jersey: Pearson Hall, 2006

Laseau, Paul *Freehand Sketching, an Introduction.* New York W.W. Norton, 2004

McGarry, Richard M., *Tracing file for Interior and Architectural Rendering,* Van Nostrand Reinhold, New York, 1988.

Mitton, Maureen *Interior Design Visual Presentation, A Guide to Graphics, Models, and Presentation Techniques.* New York John Wiley & Sons, Inc. 2008

Montague, John *Basic Perspective Drawing, A Visual Guide* New York John Wiley & Sons, Inc. 2005

Pile, John, *Perspective for Interior Designers, Simplified Techniques for Geometric and Freehand Drawing,* Watson-Guptill Publications 1985

Powell, William F., *Perspective. An essential guide featuring basic principles, advanced techniques, and practical applications,* Walter foster Publishing, Inc., 2012

Web Resources

BLOG SITES

Stephanie Sipp

Drawing + Hand, Exploring Creativity and Visual Communication
http://sippdrawing.com/

Frances D.K. Ching

Seeing. Thinking. Drawing, Drawing thoughts and observations
 http://www.frankching.com/wordpress/

Gabriel Campanario

Show and Tell, Random notes and commentary on illustration, urban sketching, art and visual communication by newspaper artist and Urban Sketchers founder Gabriel Campanario
http://gabicampanario.blogspot.com

WEBSITES

Connie Riik, Interior Designer

CSR Interiors
http://csrinteriorsinc.com/

Erica Islas, Interior Designer

EMI Interior Design
http://www.emiinteriordesign.com/

Luigi Caputo, Photographer

http://www.caputo.at/portfolio-
architecture.pho

Abstract pattern: A motif that is altered or simplified but is still recognizable.

All-over pattern: A predictable motif.

Balance: A visually pleasing distribution of objects within a composition. Balance can be symmetrical or asymmetrical.

Center line: The line that extends through the center of a cylinder.

Circle: A two-

Composition: The arrangement of objects and features within a drawing. There are several principles of design that help to create a successful composition. For example, begin with at least three objects in your drawings, and as you add more, aim to always have an odd number.

Consistency with line: The quality of keeping a drawn line the same weight throughout.

Contour lines: Lines that precisely following the curves and planes of an object.

Contrast: A perceivable and comparable difference in drawing features.

Contrast: The difference between the darkest and lightest values.

Cylinder: An object shaped like a tube.

Ellipse: A circle drawn in perspective.

Focal point: The main point of interest in a composition. The focal point has the highest level of detail, the highest value of contrast, and the largest object.

Form: The illusion of volume or mass in a drawing.

Foreshortening:

Gradual: A slow change in steps or degrees.

Ground: A surface that an object rests on.

Guidelines: Preliminary light pencil lines that can be easily adjusted.

Guide points: Light pencil dots that provide a starting and stopping place.

Horizon line: An imaginary line at eye level in a perspective drawing.

Implied pattern: A motif that is suggested or accented to create interest in a drawing.

Line: A straight mark along a page.

Line weight: The thickness of a line. Usually, the outside lines of the object are thicker, and the inside detail lines that create texture and pattern are thinner.

Line to line: The technique of touching each line to another line.

Looking ahead: The act of keeping your eye focused on the next dot or line as you draw.

Motif: A combinations of forms or shapes, as in pattern.

Major axis: A line that extends through the widest point of an ellipse.

Minor axis: A line that extends through narrowest point of an ellipse and at a 90-degree angle to the major axis.

Negative space: A background area without content that supports the overall composition.

Overlapping: The act of layering objects to show depth and distance.

Parallel: Lines that remain the same distant apart and never meet.

Pattern: A motif that creates an orderly whole.

Practice: The repetition of activities to learn a new skill.

Random pattern: An unpredictable or scattered motif. The opposite of all-over pattern.

Range: An incremental scale ranging from the darkest value to the lightest value.

Relativity: The degree of comparison between the lightest value and the darkest value of an object.

Perspective: The appearance of a distant object in relation to a viewer's distance from it.

Perpendicular: Lines that create a right angle (90 degrees) when they meet.

Proportion: The correct size relationship among the parts of an object.

Rendering: An artistic portrayal of an object

Scale: A drawing ratio that represents the true size of an object. The most common architectural scales used are ¼ inch equals 1 foot and ½ inch equals 1 foot.

Shade: A darker value on an area of an object as a result of being farther away from a light source.

Shadow: A vague image cast by an object that is depicted with a darker value.

Shape: An image in space.

Tracing paper overlay: Transparent paper placed on top of a drawing in order to retrace or refine the drawing.

Texture: The surface quality of an object.

Value: A term used to describe the lightness or darkness of an image within a range.

Vanishing point: The point where all perspective lines converge in a perspective drawing.

Variety: A difference in shape, texture, and size within a composition to add interest.

Visual weight: When comparing items that are solid and larger than a physically smaller and darker item these can appear to be same size.